On the Way to Cipango

John Cabot's Voyage of 1498

John Parsons

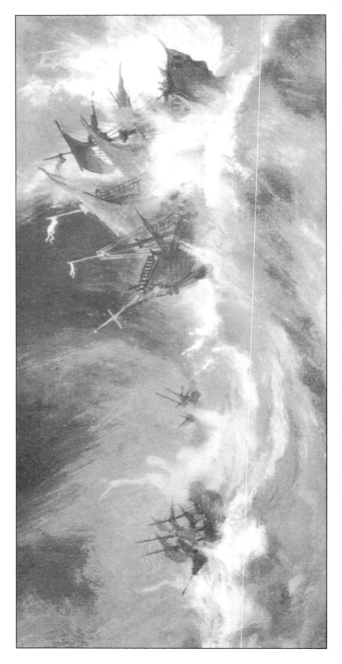

A painting by a nineteenth century artist depicting a storm at sea. Samuel E. Morison believed that Cabot's expedition of 1498 was wiped our by such a storm.

— Courtesy of Human and Rousseau Publishers, Cape Town, South Africa

On the Way to Cipango

John Cabot's Voyage of 1498

John Parsons

Creative Publishers
St. Johns, Newfoundland
1998

THE CANADA COUNCIL | LE CONSEIL DES ARTS
FOR THE ARTS | DU CANADA
SINCE 1957 | DEPUIS 1957

We acknowledge the support of The Canada Council for the Arts
for our publishing program.
We acknowledge the financial support of the Department of
Canadian Heritage for our publishing program.

The cover photo of John Cabot's statue at Bristol, England, was
taken by the author's wife, Elsie L. Parsons on April 18, 1998.

∝ Printed on acid-free paper

Published by
CREATIVE BOOK PUBLISHING
a division of 10366 Newfoundland Limited
a Robinson-Blackmore Printing & Publishing associated company
P.O. Box 8660, St. John's, Newfoundland A1B 3T7
Printed in Canada by:
ROBINSON-BLACKMORE PRINTING & PUBLISHING

Canadian Cataloguing in Publication Data

Parsons, John, 1939-

On the way to Cipango

Includes bibliographical references.

ISBN 1-894294-01-7

1. Cabot, John, d. 1498?   2. America — Discovery and exploration — British.   I. Title.

E129.C1P377 1998   970.01'7'092   C98-950262-7

DEDICATION

*This book is respectfully dedicated
to the memory of*

*Eric Martin Gosse
October 3, 1912 - December 6, 1997*

*William James Seymour
January 23, 1911 - December 30, 1997*

and

*Cecil John Clarke
July 24, 1916 - May 9, 1995*

Time, like an ever-rolling stream,
Bears all its sons away;
They fly, forgotten, as a dream
Dies at the opening day.
Isaac Watts, 1719

Unable are the loved to die —
For love is immortality.
Emily Dickinson, 1864

Our real friends do not die,
For so long as
We, ourselves, live
And remember,
They will always
Be with us
In our thoughts
And in our hearts,
An invisible
But inspiring part
Of our daily lives.
John Parsons, 1997

Raleigh A. Skelton, British cartographer, believed that La Cosa's map referred to the Cabot voyages, especially the expedition of 1498. Skelton in 1966 wrote the "Cabot" article for the *Dictionary of Canadian Biography*.
— Courtesy of the British Library, London, England.

Table of Contents

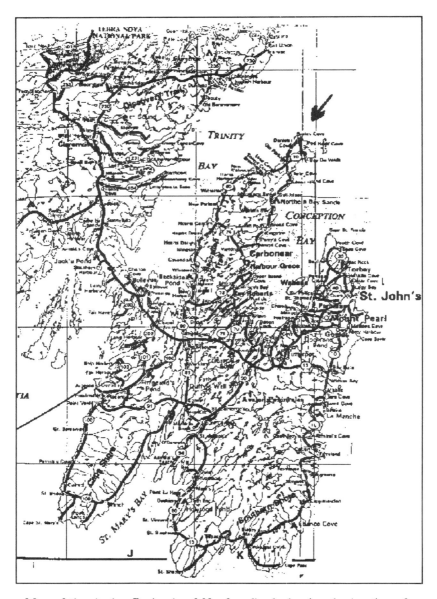

Map of the Avalon Peninsula of Newfoundland, showing the location of Grates Cove.

— The author's collection.

Foreword

Jessie Parsons

On Good Friday 1939 my husband, George, and I became the proud parents of a baby boy. Needless to say our celebrations that day turned from the spiritual to the temporal very quickly. A couple of weeks later we took our baby to St. Mark's Anglican Church in Shearstown and had him baptised. He was named John in honour of his grandfather's brother who died in 1915 while serving with the British Royal Navy. He was also named for his great-grandfather who died in 1925. In 1948 we were blessed with a girl who was named Della after one of her grandmothers. Della became a businesswoman and John became a bookworm.

John spent over thirty years as a teacher in Newfoundland and spent eight years at university. He was from elementary school always concerned with history, literature, and religion and every other scrap of knowledge contained between the covers of books. In fact, I think that there were times in school and probably university when he set out his own course of studies, and only did his required work as a sideline. Often when in high school he would write his own poems rather than try to analyse the poetry set by the teacher. In fact, John has written hundreds of poems but only a few have been published.

I am supposed to be writing this foreword about John's interest in John Cabot, but I shall get to that presently; suffice it to say, that I have always been interested in my son's literary and historical pursuits. Over the years I have been very much aware of the fact, that John has tried (sometimes with some difficulty) to search out and associate with people who can relate to his own literary and intellectual interests. One such friend was the late Eric Martin Gosse of Spaniards' Bay who died in late 1997. The two of them were intellectually compatible, despite a difference in age of twenty-seven years. John has still not completely come to terms with Mr. Gosse's death.

Little did I know almost sixty years ago when John was a baby that I would spend my last few years surrounded by books, books, and more books. Although he lives in St. John's where in his house can be found thousands of books, there are just as many or more at my house in Shearstown. Hardly a day goes by without books arriving from all over the world — from scholars as far away as London, Washington, Toronto, Miami and Vancouver. Reading and writing are passionate concerns for my son, and we, his family, have had no choice but to learn to live with it. His father and grandfather were both avid readers; in fact, his grandfather could sit down and read a book even if a war was raging behind the house.

My son has written five books, many articles and essays, and hundreds of poems, but for the past few years he has been concerned with the explorer, John Cabot who he claims did not land in New-foundland in 1497. This is essentially what his first Cabot book, *Away Beyond the Virgin Rocks - A Tribute to John Cabot* (1997) is all about. After I read this book I thought that there was very little more that could be written about Cabot. Then around Christmas 1997 came the announcement from John that he just had to research and write about Cabot's voyage of 1498. He wrote the book during January, February and March of 1998.

Now that the book is written, the major documents arranged and presented and the appendices set forth it is easy to see why the voyage of 1498 is worthy of a book in itself. Here we have the two main possible explanations as to what happened to John Cabot on his second expedition although, as John makes crystal clear in this book, few historians take the Grates Cove-Cabot Rock story seriously. In this book John presents the best interpretations of the most reputable Cabotian historians of modern times who for the most part have concluded that John Cabot's expedition of 1498 was eliminated by Spanish pirates.

The following is a quotation from Karl Samuelson's book entitled *Fourteen Men Who Figured Prominently in the Story of Newfoundland and Labrador* (1984). Here is how Karl Samuelson began his essay on Cabot in 1984:

> Considering the importance of John Cabot as the first discoverer of North America since the Norse voyages to Newfoundland and Labrador more than four centuries earlier, it is surprising how little we actually know about him. Historical research into his life has been in-tensive during the last one hundred years. But the results have left much to be desired. The current literature is largely based on hearsay evidence and by people who have made extravagant guesses. One has to carefully discard portions of information just to make sense of the rest. Only then, can we emerge with a true account of Cabot and his voyages.

The fact that Karl Samuelson includes John Cabot among the fourteen men who figured prominently in the history of New-foundland and Labrador indicates that he considered Cabot one of our important historical figures. Yet, my son, John, has gone to great lengths to prove that John Cabot was never close to Newfoundland and in my opinion, he has done a thorough job of researching the whole Cabot story. As well he has examined the writings of the major Cabotian scholars over the years, and has concluded that Cabot did not land in Newfoundland or Labrador in either 1497 or 1498. Karl Samuelson says that, "The current literature is largely based on hear-say evidence and by people who have made extravagant guesses." As can be seen by the two books John has written on Cabot there is con-siderable documented evidence in England, France and Spain to support the interpretations of the modern Cabotian scholars John has relied on to help him formulate his own judgements about Cabot. When a historian studies the original documents himself and then reads the interpretations of other historians it pretty well comes down to a matter of agreeing or disagreeing with someone else's judgement whether that judgement was made yesterday or a hundred years ago. I think that John with his expertise can determine almost at a glance what can be considered a logical historical fact, and what can be dismissed as mere speculation or guesswork.

John and his wife Elsie visited Bristol, England in April 1998. They found no visible evidence there that John Cabot ever landed in Newfoundland. The writing on the Cabot statue on the Bristol waterfront states clearly that in 1497 John Cabot discovered North America, and the east coast of North America from Key West, Florida to Cape Chidley, Labrador takes in thousands of miles of coastline. The arguments for a landfall in Newfoundland are weak in-deed; nevertheless, we did have the big celebration in 1997 including especially the voyage of the *Matthew* from Bristol to Bonavista and to seventeen other ports of call around Newfoundland and Labrador. John and his wife, Elsie, were involved in this tourist-oriented histori-cal pageant when the replica landed at Bonavista on June 24, and they were there when she departed from Trinity on August 10. They visited eleven of the ports of call and had lots of fun; yet at the time, it was my opinion that they almost wore themselves out chasing that old boat. I must confess that I enjoyed it all too. I sat in my armchair in my cosy living room and watched it all on television. That was as close as I came to the *Matthew*.

Finally, I wish to make the following comments. In outport New-foundland when I was a child in the 1920's we were lucky to survive, let alone acquire much formal education; I did, however, learn to read and write with ease, but I am not a highly-educated person like my children and grandchildren. I will admit that my daughter, Della, helped me with this Foreword which John insisted I write for his book. Of course, if this ever goes to print, I feel certain that John will give it the benefit of his editorial pen as well.

Jessie (Seymour) Parsons, John Parsons' mother and the second daughter of James and Elizabeth (Copley) Seymour was born in "The Gully" two or three years before the community was officially named Butlerville. In 1938 she married George W. Parsons and they began their married life together in the Grassy Pond area of Shearstown. Her husband died in 1973. Mrs. Parsons who has visited the Canadian mainland several times particularly Toronto and area, Prince Edward Island, and the Moncton area of New Brunswick, now alternates her time between reading, watching television and knitting; the latter activity she can do well even while watching the soaps. She also enjoys playing cards. A life-long and dedicated parishioner of St. Mark's Anglican Church, Shearstown, she is now in her eightieth year although nobody including herself wants to believe that.

Arthur Davis, who retired from Exeter University in 1970, wrote an article in 1955 in which he claimed that John Cabot died near Grates Cove, Newfoundland, during the winter of 1498.
— Courtesy Exeter University, Exeter, Devon, England

Author's Preface

Like the Preface to my first Cabot book (1997), I wish this to be regarded as a kind of personal letter to my readers as I attempt to give a little background information on the writing of this book. This short account of John Cabot's second expedition of 1498 now completes my study of the famous explorer.

I have promised my family and myself that I will not get involved with Cabot again. I hope I can keep my promise. Sometimes I wish Cabot had never left Bristol, and sometimes I wish that Prowse, Howley and Munn were right and that Cabot did indeed land somewhere on the northeast coast of Newfoundland or on the coast of Labrador. Yet, every Cabot document I lay my eyes on and every book or article I read about Cabot points to the fact that it is very unlikely that John Cabot was ever close to Newfoundland. Even if he took his bearings from what we now call Cape Race as he began his return trip in August, 1497, it is conceivable that in the fog he saw very little of Newfoundland. Hence, for all intents and purposes there is really no connection between Newfoundland and John Cabot.

Cabot's second expedition which left Bristol in May 1498 was probably more historically significant than the first voyage. Cabot was not interested in discovering new lands and especially not interested in fish. His main objective was the spices and jewels of the Orient which he believed could make him a rich man almost overnight. The presence of the continents in the Western Ocean meant that his dream could never be realized; yet, he probably died still believing that opulence lay just around the next cape.

In a way, the Foreword by my mother and the scholarly Introduction by Gerald W. Andrews have said it all, and effectively have left me kind of speechless or wordless; nevertheless, I have to tell my readers a few things about the writing of this book which I have entitled, *On the Way to Cipango - John Cabot's Voyage of 1498*.

Most books about Cabot are concerned with his first voyage and his landfall. Books written about one hundred years ago and before said very little about the voyage of 1498. Hence, scholars like Harrisse, Biggar, and Beazley have very little to say about that voyage. By James Williamson's time (1929) historians had concluded that Cabot's second voyage did indeed have much historical significance and Williamson deals with that voyage in both his books in 1929 and 1962. In recent times Ian Wilson (1991) and Peter Firstbrook (1997) have researched and written about the second voyage. This is the first complete book about that voyage.

To do my first Cabot book (1997) I had to research the second voyage if for no other reason than to satisfy my curiosity. So after I finished my first Cabot book, I decided to do an article for some magazine on the second voyage, but I soon learned as I delved into the subject that what was needed was another book, and so before long I found myself researching and writing about Cabot's voyage of 1498. The story about the Spanish pirate, Hojeda, fascinated me and I was intrigued by the fact that the great American historian, Samuel Eliot Morison, a well-known Cabotian scholar in his own cantankerous way, would not give any credence to the theory that John Cabot's expedition of 1498 was eliminated by the Spaniards on the northern coast of South America.

Then there is the Grates Cove-Cabot Rock story, a fantastic tall tale without an iota of evidence to support it. Still the people in and around Grates Cove, Newfoundland have over the years really tried to make something big out of this legend. A tall tale exaggerated by the local people is one thing, but then in 1955 a prominent geographer in England, Arthur Davies, wrote and published an article which gave a kind of international credence to the Cabot Rock story.

Much has been written about the so-called Cabot Rock at Grates Cove; yet, for all intents and purposes, the Newfoundland historian, William A. Munn, settled the issue in 1905 as regards the writings or markings on the rock. Still even today reputable scholars like Barbara Shaw, Fred Cram and Arthur Sullivan still treat the legend as if it were a fact. I hope I have laid this matter to rest once and for all. The Cabot Rock story is great stuff for tourist literature which is nothing more than a kind of legalized prostitution of history and that is about it.

The documents, appendices and the bibliography in this book speak for themselves. The documents present the evidence for the interpretations set forth, and the appendices which are secondary sources of information relate directly to the interpretations gleaned from the documents. The bibliography is a list of all the secondary sources related to Cabot's two voyages. Anybody doing a study of John Cabot should read my two books first, and then read the following books in this order: Wilson (1991), Firstbrook (1997), Williamson (1962), Pope (1997), and Cuthbertson (1997). After that read these articles as follows: Ferguson (1953), Gilchrist (1985), Jackson (1963), Kermode (1996), Hubbard (1973) and ODea (1971). After all this has been done then read Kay Hill (1968) and Jack Dodd (1974). These last two will be your reward for reading the other material. Finally read slowly and carefully R.A. Skelton's essay on Cabot in the *Dictionary of Canadian Biography*, and Robert H. Fuson's brilliant essay found in the book entitled, *Essays on the History of North American Discovery and Exploration* (1988). By that time you will have "educated" yourself about Cabot and be well-equipped to write your own book about the intrepid explorer.

I wish to express my thanks and appreciation to all those who have helped me while working on this book. These include the staff in the map room of the British Museum in London, the staff of the National Archives in Ottawa, the staff of the Naval Museum in Madrid, Spain, the staff in the library at the University of Toronto, the staff at the Metropolitan Toronto Reference Library, the staff at the Centre for Newfoundland Studies at Memorial University of Newfoundland and the A.C. Hunter Library at the Arts and Culture Centre in St. John's, Chris Caseldine of the geography department at the University of Exeter in Devon, England, Ian Wilson in London and especially Dennis Reinhartz, professor of history at the University of Texas at Arlington.

Special thanks are also due to Barbara Derksen and George Morgan in Newfoundland for their contribution to this book and to Johnny Barkerhouse and Maurice Schofield in Bristol; also, to my mother who wrote the Foreword and to Gerald W. Andrews for writing the Introduction. I wish also to express my thanks and appreciation to Don Morgan, my publisher, for his help with my books. I thank sincerely the members of my family for their love and encouragement and for their mini-grants towards my travel and research. Elsie, Cathy, Connie, John, Jessie, Della, Mark and Don don't always see the rhyme or reason in my obsession with historical subjects, but they have for the most part learned to adjust. My wife, Elsie, did an excellent job getting me around London and especially taking me to Bristol. I felt like a pilgrim as I walked upon the same ground that John Cabot walked on five hundred years ago.

Elsie and I found nothing in Bristol to connect John Cabot with Newfoundland. The statue of Cabot on the Bristol waterfront says that John Cabot discovered North America in 1497. There is also a plaque on a bridge nearby placed there by the Government of Nova Scotia in 1932 which claims that Cabot landed at Cape North on Cape Breton Island in 1497. I have heard that the erection of this plaque was a cause of considerable distress for Newfoundland historian, George R.F. Prowse. Then after Prowse survived this so-called insult, the Government of Nova Scotia designated the road around the Cape Breton Highlands as the Cabot Trail. According to Fabian ODea the naming of that road in 1941 took the "good" out of Prowse, and he died a few years later still waiting for the real proof that John Cabot landed at Bonavista in 1497.

I wish to express my sincere thanks to the History Foundation of the Hudson's Bay Company for a small grant towards my research and travel.

I wish also to thank the following individuals who have made financial contributions towards this project: Mrs. Kathleen Richardson and Dr. George T. Richardson, Winnipeg, Colonel Allan M. Ogilvie, Ottawa, Dr. Edward W. Taylor, London, England, and Mrs. Della Parsons, Shearstown, Newfoundland.

I have dedicated this book to my good friend, the late Eric Martin Gosse, my last uncle by blood, the late William James Seymour, and my father-in-law, the late Cecil John Clarke. All three in their own unique way contributed more to my education than they probably realized. They were three exceptional men, and three dedicated Newfoundlanders. I shall never forget them.

I hope and trust you enjoy this book on John Cabot's second voyage, and if you have not already done so, I hope it will provide you with the incentive to read my first Cabot book. Of course, the best approach is to read my earlier book first. Then you might not want to read this one. I hope, however, that the opposite proves to be true.

John Parsons
Shearstown and St. John's, Newfoundland
The Feast of St. Augustine of Canterbury
May 27, 1998

"The mystery of the new land." Cabot somewhere off the coast of North America. From a painting by the Canadian artist, C. W. Jefferys.
—Courtesy National Archives of Canada.

Introduction

Gerald W. Andrews

It was indeed an honour to be asked by John Parsons to write the Introduction to his most recent book, *On the Way to Cipango - John Cabot's Voyage of 1498*, his second book dealing with the role of John Cabot in the early exploration of North America. To make this latest contribution to the monumental effort he has expended over the years, to assemble the facts contained in primary sources, the analyses, interpretations and conclusions created by others concerning Cabot's two voyages, is truly rewarding. Yes, this is his second book on the subject of Cabotian studies, in addition to his other published historical works and contributions to various publications. Need I say more about his commitment and contribution to our recorded history?

In the fall of 1963, as a freshman student, I entered the main reading room of the library at Memorial University. I proceeded to a table occupied by a senior student (not John Parsons) I knew from Conception Bay. In the ensuing conversation I soon learned that he was doing a course in Newfoundland history. After a few minutes of my questioning him about his assignment, he made the following startling and almost frightening statement. "Do you know that John Cabots landfall was not in Newfoundland?"

Now imagine my situation. Here was I fresh out of the Newfoundland school system and sure that one item of definitive knowledge I had acquired was that 'John Cabot discovered Newfoundland at Bonavista in 1497' — a plain unalterable fact. To add credibility to his statement, my student acquaintance pulled from his notes an excerpt from the August 1497 Pasqualigo letter which stated Cabot passed two islands after turning back towards England. He also said that his professor, Dr. Gordon O. Rothney, believed that Cabot's landfall was on the mainland of North America somewhere southwest of Newfoundland — most likely in southern Nova Scotia or Maine.

One item in my belief system was being shattered. A small pillar of my knowledge was being taken away. I truly felt a sense of insecurity. No doubt this is how most people will feel when they begin to read John Parsons' books about John Cabot. The more aggressive souls may even feel some anger. What we believe is the major part of our reality. Be prepared to have a component of your reality changed by this book.

Some years later at the end of my career as a university student I was in the process of defending a thesis in front of a group of graduate students and professors. At the conclusion of the exercise the audience left the room, all except one professor who quietly came to me as I was folding my papers and said, "Always remember Gerald that all knowledge is tentative." This statement has again been solidly reinforced to me by the scholarship of John Parsons in his two Cabot books.

The quest for knowledge is essentially a search for truth. All humans are consciously or unconsciously engaged continually in the process. Police detectives, scientists, philosophers, historians and other scholars apply very formal means to the search for knowledge — technically it's called epistemology. In my estimation, John Parsons has applied a sound epistemological process to history in his Cabot books. He has assembled the facts - the documentary primary sources and the cartography. He has assembled the secondary sources - the letters and statements of Cabot's contemporaries, and he has presented conclusions and judgments reached by numerous professional and amateur historians, statements from members of the general public and the media who have attempted to interpret those sources. To organize this storehouse of knowledge he has included extensive appendices in both books and recently compiled a separate extensive and up-to-date bibliography of sources in the area of Cabotian Studies. He has done us a great service by presenting us with all this material so that we too can engage in the enigmas of John Cabot — the compelling unanswered questions about him. In addition to the general readership, I recommend this book to professional students and teachers of history who must develop the intellectual tools of the historian.

In these works, John has gone to great lengths to separate facts from legends, knowledge from speculation, myths from celebrations and the stuff of tourism promotions. He has attempted to create the pure history. But history is far more than a compilation of facts. It includes the human interpretation, understanding and responses to the events. John has certainly played in this arena as well. He has given his interpretation to the facts — it was his responsibility to do so. In so doing he has created some controversy which is always a good catalyst for further intellectual pursuit. But he has given us the grist for our own thought mill, the ammunition to disagree if we choose a different interpretation of the facts and events. He has been truly honest with his readers.

The most significant components of the Cabot story are not so much the exact location of his landfall, or how far he proceeded on his second voyage, or even where he perished, but what happened subsequently as a result of his voyages of exploration. This book challenges us to further examine the role of John Cabot in the major North American historical developments that followed his lifetime. Did John Cabot actually have a greater role than Christopher Columbus in the initial development of the current North American society? The breadth of knowledge supplied by John Parsons in these books, and particularly the latter, may force us to vastly expand our concept of Cabot's contribution in this area. These works by John Parsons are indeed a "Tribute to John Cabot."

Some months ago before I had seen this manuscript, the author mentioned that John Cabot's second voyage in 1498 was probably more historically significant than his first voyage a year earlier. After reading this book I now understand why that statement is tenable.

In closing, I want to remind readers that up to the middle of the last century Sebastian Cabot, not John Cabot, apparently through deception, was credited with discovering the new lands in the Western Ocean for the King of England. Newfoundland historians up to the late 1800's, before Prowse, were stating that Harbour Grace, not Cupids, was the original site of John Guy's colony in Newfoundland. Up to the 1960's, most North American historians would not credit Newfoundland as the site of the Viking settlement as described in the Norse Sagas. Archeological evidence proved otherwise. Fortunately our true historical knowledge keeps unfolding and evolving.

The story of John Cabot is a significant part of North American history. We owe a great debt in recent years to all those Cabotian scholars including especially John Parsons who have worked so hard to further illuminate that aspect of our history for us.

Gerald W. Andrews (b. 1946), is a native of Port de Grave, Newfoundland. He graduated from St. Lukes High School, Port de Grave, and Memorial University of Newfoundland and holds a Masters degree from the University of New Brunswick. He worked one year as a mathematics instructor in the Community College System, eighteen years as a

school teacher, and 10 years as a Program Coordinator in the secondary school system. Since 1971, Gerald has lived at Carbonear with his wife Rose (nee Dean) where they have raised four children. He retired from the education profession in 1997.

Since his undergraduate years at university, he has maintained an active interest in the study of Newfoundland history and culture. He was the founding president of the Port de Grave Peninsula Heritage Society in 1993. In 1997 he published his first book, Heritage of a Newfoundland Outport: The Story of Port de Grave, a comprehensive history of his home community. This book has been well received not only in Newfoundland but in other parts of North America as well.

King Henry VII by an unknown contemporary artist. This monarch, who died in April 1509, has been called "England's first modern king."
— Courtesy National Maritime Museum, Greenwich, England

Sebastian Cabot, son of John Cabot, confused historians for three hundred
years about the Cabot expeditions. This protrait is a copy of a painting by
Hans Holbein.

— Courtesy of the New York Historical Association,
and the National Gallery, London

Prologue

George Parker Winship (1871-1952), an American historian, is well-known in the field of Cabotian studies for his *Cabot Bibliography* (1900) in which he lists 579 items about John Cabot and his son, Sebastian. In the Annual Report of the American Historical Association for 1897, published in 1898, this historian wrote a short but informative essay about Cabot entitled, "John Cabot and the Study of Sources." Some of the things Winship wrote in his essay in 1898 are just as applicable in 1998 as they were one hundred years ago. Here are some extracts from that essay:

> The earliest printed reference to the discovery made by Cabot was published half a century after the date of the voyage. Some years earlier, in 1516, Peter Martyr published an account of a voyage by a Cabot, and this account was reprinted, circulated widely, and was frequently copied. Other accounts, giving various details of a voyage by Cabot to the north and west, were published by Ramusio, an Italian correspondent of Sebastian Cabot; by Richard Eden, who knew the younger Cabot intimately; by Gomara, Galvano, Oviedo, and by others who were not only contemporary with Sebastian Cabot, but who lived in the same places and moved in the same professional circles with him. During the second half of the sixteenth century the English chroniclers, Grafton, Holinshed Fabyan and Stow, Hakluyt and Herrera published accounts of the Cabot voyage, several of which contain statements that do not occur elsewhere. All of these writers were well acquainted with men who had been associated with the younger Cabot. The books which they published are the authority for a large part of what has been written about the periods of which they treat.
>
> The statements in these printed books often differ materially from one another. Not one of the writers describes more than a single voyage by Cabot to the northwest, and the descriptions given are often mutually impossible. Not one of them reports that Sebastian ever spoke of any voyage made by his father. Hence it has been deduced that Sebastian was a braggart and a liar, who persistently strove to secure for himself the credit of his father's achievements. And therefore, to complete the argument, it is stated that Sebastian never achieved anything of importance by himself, and that he was

1

not competent to accomplish anything.

The direct connection with the Cabots ceases after 1600. For the next two hundred years their discovery is frequently mentioned by succeeding voyagers, by historians, and by sermonizers. Occasionally one of these ventured to draw some inference from the confusion of the earlier writers, but the impression which this confusion made upon students and the public was fairly stated by Burke in 1757, when he wrote: "We (English) derive our rights in America from the discovery of Sebastian Cabot, . . . but the particulars are not known distinctly enough to encourage me to enter into the details of his voyage. . . ."

The confused tangle which had grown out of the earlier printed narratives has been cleared away by the finding of manuscript sources, recovered from the storehouses of documentary material. The first of these sources was made known by Richard Biddle, a Pittsburgh lawyer, who printed in 1831 a document which proved that there had been two Cabot voyages of discovery. A few years later Rawdon Brown found in Venice a letter written from London in 1497, which describes the effect produced by the return of Cabot in August of that year. Rawdon Brown in Italy, and Bergenroth in Spain, carried on the search for historic manuscript material, and by 1870 a half dozen letters and official reports had been found, dated in 1497 and 1498, in which Cabot is mentioned, and which repeat some of the current gossip about his voyages and his future plans . . .

Just as a single document found in the London record office in 1830 proved that there was no longer any need of crowding all the events of the Cabotian story into the course of a single voyage, so the finding of an old map in the library of a Bavarian curate in 1843 gave us a direct statement, apparently made by Sebastian Cabot himself in 1544, crediting his father with the discovery of North America. Similarly, within a few months, some memoranda of the customs collectors in old Bristol have been found among the Westminster muniments, which are said to prove that John Cabot was in England in the autumn of 1498 or 1499. Before the discovery of this manuscript no mention of the existence of John Cabot after the departure of the expedition in the spring of 1498 had been known. Hence it had been inferred by nearly every writer upon the Cabots that the father died before that expedition returned, so that all the glory of that voyage descended upon his son. If John Cabot was alive in England in 1499, a considerable portion of all that has been written about the Cabots loses its value as a statement of truth, but its value is correspondingly increased to the student of how history is made.

Mr. Biddle found the explanation of the printed accounts in the manuscript sources. He was also the first to interpret these sources of Cabotian history; to erect inferential structures out of the presumptions which might be drawn from these sources. As it seems to me, the most important portion of the whole body of

Cabotian literature is that which reveals the mental processes by which the eulogists and the detractors of Sebastian Cabot have reached their conclusions . . .

Ramusio in Venice printed his recollections of what he had once heard at a house-party in Verona, where a chance acquaintance told of a conversation with Sebastian Cabot many years before in Seville. Ramusio's narration does not mention John Cabot; hence, argues one authority, Sebastian was guilty of unfilial falsehood. Henry VII gave John Cabot a charter in March, 1496, and seventeen months later John Cabot returned to London. Bristol ships traded to Iceland, and therefore, says an Oxford investigator, Cabot spent the winter of 1496-97 in Iceland. Somewhere it is stated the Cabot landfall was 50^0 north latitude. Hence a right reverend bishop declares his belief that Cabot first saw the soil of North America at Cape St. John, across which runs the line of 50^0 north, according to the perfected instruments of 1897.

And much more of the same sort of argument from the honest essays of men, each of whom fairly deserves the serious respect and consideration of fellow-students — much more of equal interest to us who believe that an historian ought, first of all, to possess common sense and some appreciation of how men and women are likely to act and think.

I want to plead for the study of the Cabot question, not by you, college teachers, whose historical training and developed instincts might be so much more usefully employed, but by the scores of young men and women who come to you, anxious to study history, filled with enthusiasm for the subject and confident of their graduated ability to understand what older men and women have done and are doing. I wish that every would-be historian could begin his professional training by preparing for an examination on what has been known and what might be known about John and Sebastian Cabot. The history written afterwards would be marked less often than now by blind quotation from the Sources, and less by illogical conclusions maintained by baseless inferences and unwarranted assumptions. [1]

What is the substance of what Winship wrote in his essay in 1898 which can be gleaned from these extracts? Effectively, he is saying that up to his time much that had been written about John and Sebastian Cabot was based on the secondary sources produced in the hundred years or so following the Cabot voyages. He is also saying that much of this information is unreliable, and that only after Richard Biddle's time (1831) were serious attempts made to write valid historical accounts of the Cabot voyages based on the contemporary documents, both literary and cartographical, discovered up to that time.

Between 1880 and 1930 valid or reliable accounts were written by well known Cabotian scholars such as Henry Harrisse, Justin Winsor, Charles Deane, Henry P. Biggar and James A. Williamson. Harrisse's best known book (1896) has been called, "a laboratory manual, in which the student finds revealed each step of the processes through which the material of history has been forced in order that it might be made to render up the truth which was contained within it," while Charles Deane's essay in Justin Winsor's major study *Narrative and Critical History of America* (1884) was labelled by Winship as, "a comprehensive survey of the Cabot sources and secondary authorities." [2] Henry P. Biggar, the Canadian historian, in books published early in this century made an attempt to improve on all that had been written about Cabot's voyages up to that time, and James A. Williamson's book (1929) was a serious attempt by a serious scholar to do a thorough job on the Cabot voyages by making full use of all the primary and secondary sources available. [3] The only important Cabot document not available to Williamson in 1929 was the John Day letter discovered in 1956. Williamson discussed this document in his classic study in 1962.

Up to Williamson's time most of the emphasis in the secondary sources was on Cabot's first voyage with very little attention paid to his second voyage. This was understandable since not much was known about Cabot's ill-fated voyage of 1498 from which he did not return. Still the second voyage in some ways was more significant than the first. The first voyage, as Williamson has pointed out in both his books, was a voyage of reconnaissance whereas the second voyage was a serious attempt to reach the riches of the East and somewhere along the route establish a colony which would serve as a stepping- stone or half-way settlement between the East and London which place it was hoped would become so involved in the spice trade that it would rival Alexandria in Egypt, the half-way trading centre between Europe and Asia. [4] Williamson recognized this point in his first Cabot book (1929) and gave due attention to it.

> The whole fleet, according to De Ayala, was provisional for a year; and since he added that they hope to be back by September, it may be that the provisioning was for a colony to remain in the new land . . . The purpose of the voyage [i.e. 1498] was . . . to follow the new coast to tropical latitudes and thence to discover the spice regions adjoining Cipango . . . A project for a colony would seem to be rather out of place in opening relations with the powerful princes who were supposed to rule over the regions sought for. It is possible the settlement was intended to be on the coast founded in 1497, where it would have been useful as a half-way depot at which

goods from the spice islands could be collected and thence forwarded to England. [5]

Arthur P. Newton, a British scholar, in his book published in London in 1932, only three years after Williamson's study on Cabot, claimed that Cabot returned to England from his second voyage. Newton assumes that Cabot and his men, because their expedition was a commercial failure, did not get a very good reception from their backers: "Of the reception accorded to them on their return we know nothing, but it is not difficult to understand that it was not enthusiastic. Instead of bringing spices and oriental gems, their ships were empty of everything, and those who had supplied money for the expedition realised that it was lost for good and all." [6] Newton's book is based on a series of lectures he delivered at King's College, London, in 1931 and one would think that these lectures would contain reliable historical information, but that was not the case. Newton, a professor of history, believed that Cabot's expedition returned, and furthermore he was of the opinion that Sebastian was on the 1498 voyage. Newton, a so-called distinguished scholar, cannot be forgiven for disseminating incorrect information. Surely, Williamson's well-documented study published only three years before was available to him.

Robert H. Fuson, a modern-day American scholar, writing in the 1980's believes that there is still a lot about the Cabot voyages we do not know, even though he concedes that nowadays historians take special pains when publishing results of their research to make certain that their information is reliable. To Fuson, "John Cabot was, and is, an enigma," a man whose greatness as an explorer was almost stolen by his son Sebastian. Here are two extracts from Robert H. Fuson's brilliant essay on John Cabot:

> One hundred four years have elapsed since Henry Harrisse set out to correct the historical record and to restore John to his proper place among fifteenth-century navigators. Dozens of books and articles have appeared during the past century as scholars sorted through the old records and pored over the early nautical charts. The last significant find came as recently as 1956, when Dr. L. A. Vigneras discovered the John Day letter in the Archivo General de Simancas. Nevertheless, in spite of all of this attention to the Cabots, our ignorance is appalling.
>
> We do not know where John Cabot was born, when he was born, or even his exact name. There were no contemporary portraits or physical descriptions that we know of, so his appearance is a mystery. No extant document informs us of his

residency before he went to Venice. Only fragments of his family life have come down through the years; his father's name is in doubt, and nothing is known about his mother, two of his sons, or his brother . . .

John made a short unsuccessful voyage in 1496, but all details of it remain a secret. There is no absolute date for the departure or return of the 1497 voyage, and the North American landing site has never been determined.

Lastly, we do not know when John died, where he died, or how he died. His death may have been at sea or anywhere on land between Canada and Florida.

Such is the stuff that makes good historical fiction, if not good history. John Cabot, unfortunately for us, was born into a world that did not keep very good records for persons of humble origin. Biographies of such men came later, after the accidents of history and geography thrust fame and fortune upon them. But John's untimely death permitted little of either. John Cabot himself left not a single holograph scrap - not a letter, not a map, not a ship's log. Nothing. Nothing, that is, except Sebastian. [7]

Fuson's statement that John Cabot left nothing except Sebastian is a significant one for had he not left Sebastian, then maybe there would not have been so much confusion about the two Cabot voyages. Sebastian, in relating information to historians of his time, either told deliberate lies or had such a bad memory that he just could not get anything straight. He seemed to be confused, and his thoughts totally mixed up. He informed historian Peter Martyr, the Italian historian, who spent most of his life in Spain, that he was the commander of the second voyage, and that the expedition returned to Bristol. Sebastian's description of the second voyage was in reality a description of his own voyage made to the northwest in the years 1508-1509. In fact, Sebastian was on that voyage in April 1509 when King Henry VII died. [8]

Fuson's essay emphasizes the point that John Cabot was not really interested in discovering new lands. He was interested in trade with the East. This was his objective from the beginning and his first voyage in 1497 was merely an attempt to convince himself that the East lay towards the west across the Atlantic Ocean. He did indeed find land in the western ocean, but it was a long way from the Orient. Still, once he found land in 1497, his objective was accomplished and his mind set on a more significant voyage the next year in which he hoped to make contact with the people of Cipango and Cathay and begin serious trading operations. The following extract from Fuson's essay is significant in that it relates directly to Cabot's motivation:

John Cabot may have gone to England in 1494 or early 1495. He probably went first to London, then to Bristol. Here he learned that the Men of Bristol had already done what Columbus had done — failed to journey far enough. But the Men of Bristol were not looking for Asia. They may have been seeking the legendary Isle of Brazil or the Island of the Seven Cities, or perhaps nothing more than a good fishing hole. In any event, Cabot was successful in obtaining permission to sail *beyond* Brazil and the offshore islands found by the Spaniards. Once at the mainland a turn south would fetch Cathay. [9]

So Cabot's objective right from the beginning was, "to sail *beyond* Brazil and the offshore islands found by the Spaniards," for having found the mainland in 1497, he was of the belief that, a turn south would fetch Cathay. But that was not to be, for his death in 1498 or 1499 ended his dream, and he died still thinking Cipango and Cathay were within his reach. Cabot's expedition of 1498 was well-planned and well-organized and his plan for a short route to Asia was a good one, but all was destined to meet with failure, because a gigantic land mass stretching almost from pole to pole blocked his way and crushed his dream.

We are left with the question of what happened to Cabot when his expedition of five ships faced the cold dark uninviting Atlantic Ocean in May 1498. A definitive answer to that question cannot be given in 1998 five hundred years after he vanished from this earth forever, but we do have sufficient information based on both primary and secondary source material to make an attempt at answering that important question.

There are two theories that can be put forth. One is that he died a violent death on the northern shores of South America later known to historians as the Spanish Main; the other is that he died an even more terrible death on the rocky coast of Newfoundland near the present-day settlement of Grates Cove, Trinity Bay. [10]

The arguments for and against these two propositions are presented herein and readers must draw their own conclusions; first from a careful reading of what I have written based on the primary and secondary sources, and then from a detailed analysis of the contemporary documents presented. The rest of the book might prove to be edifying as well, and for those who might wish to become the world's latest Cabotian scholars, then I recommend reading every item in my *Cabot Bibliography*, beginning with Raleigh A. Skelton's essay on John Cabot in the *Dictionary of Canadian Biography* (1966) and ending with my book, *Away Beyond the Virgin Rocks* (1997).

Notes to Prologue

1. George P. Winship, "John Cabot and the Study of Sources," *Annual Report of the American Historical Association for 1897*, (Washington, D.C. 1898).

2. *Loc. Cit.*, *Notes* accompanying the Winship essay.

3. See Henry P. Biggar, *The Precursors of Jacques Cartier 1497-1534, A Collection of Documents Relating to the Early History of the Dominion of Canada* (Ottawa, 1911), and James A. Williamson, *The Voyages of the Cabots and the English Discovery of North America under Henry VII and Henry VIII* (Cambridge, England, 1929).

4. James A. Williamson, *The Cabot Voyages and Bristol Discovery under Henry VII* (Cambridge, England 1962). pp. 114-115.

5. James A. Williamson, 1929, *op. cit.*, pp. 180-181.

6. Arthur P. Newton, *The Great Age of Discovery* (London, 1932), pp. 136-137.

7. Robert H. Fuson, "The John Cabot Mystique," in Stanley H. Palmer and Dennis Reinhartz (eds.), *Essays on the History of North American Discovery and Exploration*, Arlington, Texas, 1988, p. 35-36.

8. Williamson, 1962, *op. cit.*, Chapter IX.

9. Fuson, *op. cit.*, pp. 46-47.

10. Williamson, 1962, *op. cit.*, Chapter VII. See also Arthur Davies, "The Last Voyage of John Cabot and the Rock at Grates Cove," *Nature*, Volume 176, November, 1955, pp. 996-999 and Ian Wilson, *The Columbus Myth* (London, 1991), Chapter 10.

Chapter 1

The Voyage of 1498: An Overview

History records that an unsuccessful voyage of exploration into the Atlantic was made by one John Lloyd of Bristol in 1480. Other voyages from Bristol followed in the 1480's and early 1490's and, according to Tryggvi J. Oleson and William L. Morton, "There is evidence, though not certain proof, that one of these voyages resulted in the discovery of North America by the English before 1492." [1]

These attempts by the seamen of Bristol served as an incentive for John Cabot, the Italian navigator, who with his English crew in the caravel, *Matthew*, landed on the west side of the Bay of Fundy in June 1497. On August 6, 1497 Cabot was back in Bristol. Soon a decision was made to cross the Atlantic again in the summer of 1498. In the words of Oleson and Morton: "The next year Cabot apparently perished on a more ambitious voyage." [2] Cabot's more ambitious voyage of 1498 is the subject of this book.

Cabot's objective on the voyage of 1498 was to follow the coastline in a southwesterly direction from where he landed in 1497 until he came to the land of the Great Khan. The extensive preparations for the new expedition suggest that Cabot's intention was to establish a colony somewhere in that area, a colony that would initially be made up of prisoners supplied by the King of England. The new royal letters patent granted on February 3, 1498 authorized the expedition to use six English ships of about 200 tons each and for them to proceed, "to the lande and Iles of late founde by the seid John," and to take with him enough of the king's subjects to establish his colony. [3]

When the expedition did get underway about the beginning of May, 1498 there were five ships instead of six. One was equipped by the King and hired from two London merchants, Lancelot Thirkill and Thomas Bradley, who may or may not have gone on the voyage, while the other four were supplied and equipped by merchants of London and Bristol based in Bristol. According to a letter by the Spanish ambassador to England dated July 25, 1498 Friar Buil, an Italian priest went on the voyage and, according to a letter from Agostino de Spinula dated June 20, 1498 and sent to the Duke of Milan, a Milanese cleric of high rank (probably a bishop) named Giovanni Antonio de Carbonariis also went on the expedition. The ships were provisioned to last for up to one full year. [4]

There is no official record of Cabot's voyage of 1498 and what really happened to Cabot and his five ships and their crews can only be inferred from various pieces of information gleaned from several documents of that year and subsequent years, particularly the letter by the Spanish ambassador dated July 25, 1498. [5] Not far west of Ireland the expedition encountered a storm and one ship was so badly damaged that it had to return to an Irish port. This ship may have resumed the voyage as is suggested by a statement in a work by Polydore Vergil written and published in 1513. Vergil implies that Cabot sailed first to Ireland and then faced the Atlantic again and headed west: "In the event he is believed to have found the new lands nowhere but on the very bottom of the ocean . . . since after that voyage he was never seen again anywhere." [6]

James A. Williamson (1886-1964), a recognized authority on Cabot, believed that even if the ship Cabot was on was lost it is likely that one or more of the other four ships managed to weather the storm and reached land somewhere on the coast of North America. [7] One or more may even have returned to England, though there is no record to that effect. Raleigh A. Skelton (1906-1970) says the evidence that one or more of the ships made it across the Atlantic is "slight but cumulatively significant," and he goes on to mention in particular the piece of broken gilt sword of Italian workmanship and the two silver earrings that were of Venetian origin. [8] These items were found by a Portuguese expedition of 1501 commanded by Gaspar Corte-Real. Both Skelton and Williamson believed that these objects, which could have been found on the coast of North America anywhere from Labrador to New England, came from Cabot's voyage of 1498.

On June 8, 1501 the Spanish monarchs issued letters patent to one Alonso de Hojeda for a new voyage of exploration in the Southern Caribbean along the northern coastline of South America. Hojeda was instructed among other things to "follow that coast which you have discovered, which runs east and west . . . because it goes towards the region where it has been learned that the English were making discoveries; and that you go setting up marks with the arms of their Majesties . . . so that you may stop the exploration of the English in that direction." [9] The letter by the Spanish ambassador, Pedro de Ayala (July 25, 1498) also gives an indication that there were Englishmen in the Caribbean area and that these English adventurers were trespassing on Spanish territory. [10] Any Englishmen in that area were to be stopped by Hojeda and his expedition.

The only English explorers, if any, who were in that area between 1498 and 1501 had to be part of Cabot's expedition since there is no record of any other English expedition around that time. The five ships could have been together; on the other hand, between Bristol and the northern coast of South America some of the ships might have been lost, and hence Hojeda could have encountered only one or perhaps two ships on board one of which John Cabot may or may not have been. Hojeda was a ruthless adventurer known for his cruelty towards the natives and even his own people in that region. Without any feelings of compunction it would have been a small matter for him to engage the English explorers in a fight and proceed to eliminate them right down to the last man. [11]

That Hojeda encountered Englishmen in the Caribbean between 1498 and 1501 is a likely possibility; still Williamson says that Hojeda's letters patent of June 8, 1501 is, strickly speaking, not evidence that the Spanish explorer encountered any Englishmen in that region on his expedition in 1499. Yet the conclusion cannot be ruled out for, according to this English scholar, it is supported by a "sober and responsible" Spanish historian, Martin Fernandez de Navarrete. In 1829 Navarrete had written: "It is certain that Hojeda in his first voyage [1499] encountered certain Englishmen in the vicinity of Coquibacoa." [12] Coquibacoa is marked on some early maps as being just to the east of Caba de la Vela.

Williamson writing in 1962 claimed that, "Two commentators consulted by the present writer consider that Navarrete may have been making an unwarranted jump in advance of the evidence, a guess based on the mention of Coquibacoa in Ojeda's (i.e. Hojeda's) privileges." [13] Still most historians consider Navarrete's statement to be authentic and unqualified. *Lo cierto es*, he recorded, "It is certain." Navarrete, who does not give the source for his statement, may have been familiar with some document that has since been lost.

Williamson sums up the matter of the nineteenth century Spanish
historian this way:

> So we must leave this question tantalizingly uncertain. One
> thing may be added, that the Spanish information need not have
> been obtained through England. If Navarrete is right, it could have
> been gathered from the Englishmen encountered at Coquibacoa,
> and these men may never have reached home. The ship concerned
> may even have been John Cabot's missing flagship. Possibilities:
> nothing more. [14]

We have no real knowledge of the discoveries made by the second
Cabot expedition, although knowledge of that expedition could be
reflected in the La Cosa map of 1500, for La Cosa could very well
have incorporated into his map knowledge gleaned from both Cabot
voyages. There is no way to know for certain. The evidence inherent
in the Hojeda licence is that some southward advance was made by
some Englishmen in the years 1498 to 1500 and the only Englishmen
that could have met Hojeda's expedition of 1499 were those of
Cabot's voyage of 1498.

Nothing was heard of the second Cabot expedition by the end of
September 1498. It can be inferred from Pedro de Ayala's letter of
July 25, 1498, that the expedition was to report back to England by
that date.[15] Maybe Cabot intended to return to inform his financial
backers and King Henry VII about the progress of the trading colony
he had founded. But nobody returned. Cabot's ship may have been
lost off the northeast coast of North America in the winter of 1498-99
as he attempted to return to England. In fact, most general history
books and encyclopedias say that Cabot was lost at sea during the
winter of 1499, but this is all based on conjecture rather than any
kind of documentary evidence. [16] The whole expedition could have
made it into the Caribbean area, the Spanish sphere of influence, and
the English adventurers could have all been murdered by the
Spaniards. In any event, the last officially recorded event in which
John Cabot participated occurred when his second expedition of five
ships left Bristol in May 1498.

It was formerly believed that Cabot was in England in 1499 because his pension was paid up to the end of the financial year 1499
(Michaelmos 1498 to Michaelmos 1499). Since Cabot at the time was
not known to be dead, his pension was evidently paid to a family rep-
resentative most likely his wife, Metta. People in and around Bristol
particularly Cabot's financial backers waited in vain for news of the
expedition which did not come. James A. Williamson summed up the

matter this way:

> The lack of a focal point in the return of the expedition, and its
> failure as a commercial venture, account for its passing-over by the
> chroniclers. People waited for the return of Cabot, and he did not
> come. As the months passed they gave him up for lost. [17]

There is very little evidence as to what happened to John Cabot
and the men who sailed on this second expedition in 1498, but what
little information we do have, or even informed speculation gleaned
from documents of the time and historical writings since, must in
Williamson's metaphorical expression be regarded as "a pinhead of
gold in a desert of dross." [18] Williamson wrote two books about the
Cabot voyages: one in 1929 and the other in 1962. It would appear
that, in his 1962 book, this dean of Cabotian studies intended to do
the definitive study of Cabot, yet at the end of the Preface to that
book he wrote that, "Cabot study is never finished . . . It is a living
subject." [19]

To assist Williamson in his study of Cabot, Raleigh A. Skelton,
one of the great cartographers of the twentieth century, did a study
of the maps related to the Cabot voyages, particularly the La Cosa
map dated 1500 which was discovered in Paris in 1832. [20] Both Williamson and Skelton believed that in 1497 Cabot landed in the Bay of
Fundy area and in 1498 one or more ships of Cabot's expedition
ended up on the coast of South America in the Caribbean sea. La
Cosa was with Hojeda's Spanish expedition in 1499 and may have
benefitted from the geographical knowledge of North America and
the Caribbean that the Englishmen had gained from their explorations. How else could La Cosa's map be so exact in certain areas that
the Spanish had not explored up to that time particularly the area
west of Coquibacoa in the Gulf of Darien and the North American
coastline north of the Caribbean? [21]

This being the case then, the Englishmen, if they were murdered
by the Spaniards, obviously did not get a chance to return to England
and relay to King Henry VII the important information regarding the
discoveries they had made on the North and South American continents.

In the 1960's and 1970's Samuel Eliot Morison (1887-1976), the
well-known American historian, wrote several books about early
discovery and exploration in North and South America. Morison dismissed outright the theories of Williamson and Skelton concerning
Cabot's voyages. Morison did his own interpretation of the

documents relating to the voyages and ignored Williamson's belief that one or more of Cabot's ships may have encountered the Spaniards in the Caribbean Sea. He also disagreed with Williamson who claimed that Cabot on his first voyage in 1497 likely landed in Maine. There are times in his writings when Morison seems to be too flippant and gives the impression that he has all the answers:

> Williamson to my astonishment, gives Cabot a Maine landfall in order to fit in with his identification of the English Coast on the La Cosa map as Nova Scotia. Although second to none as a Maine-iac, I cannot accept this. It ignores Day's evidence . . . and adds many hundred miles to what he had to cover in 26 days. [22]

Morison even has a word of caution for his readers and urges them to be on their guard against certain English and Canadian historians: "The reader should keep in mind that English and Anglo-Canadian historians are desperately eager to prove that Cabot touched the American mainland, so that they can claim a 'first' for him as discoverer of the continent, Columbus not having set foot on the mainland before 1498." [23] With reference to Englishmen being in the Caribbean area in 1499 or 1500, Morison in another one of his books makes this rather sarcastic remark which no doubt was intended to wound the pride of those who endorse Williamson's theories about Cabot: "Although Ojeda [Hojeda] found no trace of English activity on this coast, this has not prevented certain 'English first' enthusiasts from claiming that John Cabot came this way on his 1498 voyage and had been observed by Spaniards." [24] Effectively then, Morison dismisses the theory which can certainly be implied from early Spanish documents that John Cabot or part of his expedition made it as far south as South America in 1499 and were exploring in that area when they encountered the Hojeda expedition. Yet a historian of Morison's calibre should have known that no scrap of historical evidence should be ignored in the process of historical interpretation and reconstruction.

In 1991 an English scholar, Ian Wilson (b.1941), published his book entitled, *The Columbus Myth*. In this book, Wilson supports the theories of Williamson and Skelton regarding Cabot's voyage in 1498 and does an excellent job of interpreting the documents. His book is very well-written and no doubt he has given the best account yet of what possibly could have happened to Cabot or part of his expedition when they met the Spaniards in the south. Like his historical mentors, Williamson and Skelton, Wilson also believes that in 1497 Cabot landed to the south of Newfoundland. In his view, if the La Cosa map is any reflection of Cabot's first voyage (and many historians think it is) then Cabot's landfall had to be somewhere on the mainland beyond Nova Scotia — most likely on the coast of Maine:

> If La Cosa's map is indeed of the 1497 Cabot voyage, then it incidentally provides the strongest possible evidence that Cabot's landfall had indeed not been Newfoundland, but further to the south along the eastern American seaboard. For impossible as it is to identify La Cosa's coastline with any exactness, it is most certainly not that of a heavily indented island such as Newfoundland. [25]

When I began my study of Cabot and started to interpret the relevant documents and then evaluate the theories of other historians, the theories of Williamson and Skelton and later Wilson seemed the most logical and certainly made the most sense to me. Hence in that sense I am a disciple of these English historians even though my writing of history is my own and not necessarily a regurgitation of theirs. Every historian after studying the relevant documents has to be influenced one way or another by the interpretations of other historians who have previously ploughed the historical fields.

I would recommend to my readers that they consult all the important works about Cabot, particularly the latest contributions to Cabotian studies by Peter E. Pope, Brian Cuthbertson, Francesco Alusio and Peter Firstbrook. And, of course, no study of Cabot is complete without a thorough study of the works of Henry P. Biggar, William F. Ganong, Henry Harrisse, Samuel E. Morison, David B. Quinn, and Samuel E. Dawson.

So what happened to John Cabot in 1498? When he cleared Dursey Head in May of that year as commander of an expedition of five ships, he sailed right into history and was never heard from again. At least there is no English record or document extant which says otherwise. As I have pointed out there is evidence but not proof that Cabot or part of his expedition may have gone too far south and hence sealed their own fate; on the other hand, he may have been shipwrecked anywhere on the coast of North America from Labrador to Florida or beyond. Recall that there were five ships and, as Williamson has pointed out, it is unlikely that they were all lost together. It is also unlikely that they were all lost separately. There is no record since the fifteenth century other than Cabot's voyage of 1498 to indicate the loss of several ships and there is no proof that all of Cabot's ships were lost at sea. Some of the five ships may have been lost in a storm at sea or shipwrecked on shore. Some may indeed have made it to the coast of South America and their crews then eliminated by the Spaniards.

There is a theory that John Cabot was lost in Baccalieu Tickle near the settlement of Grates Cove in Newfoundland. The theory is associated with the so-called "Cabot Rock" and the engravings thereon which the local people in and around Grates Cove believe were put there by Cabot or members of his crew after they were shipwrecked. Information about the "Cabot Rock" has been known since the early nineteenth century and history records that John Cabot did not return from his voyage of 1498. The logical thing then for the local people to do, especially amateur historians and local writers, was to connect Cabot's mysterious disappearance with the rock and hence the legend that has grown up around it. [26]

In 1955 a reputable English scholar Arthur Davies (b. 1906) of the University of Exeter in Devon, England wrote an interesting but far-fetched article in the well-known periodical, *Nature* magazine. [27] Davies' article, because it was written by a well-known British geographer, gave credence to the theory that Cabot was indeed lost in Baccalieu Tickle. Local people have even erected a small monument in Grates Cove and over the years they have used the legend of the Cabot Rock to best advantage particularly as regards the tourist industry. In fact, so persistent were the local people in their belief that Cabot ended his days at or near Grates Cove that the powers-that-be arranged for the *Matthew*, the replica of Cabot's ship, to stop at Grates Cove on June 27, 1997. After a brief ceremony that day a wreath was dropped in Baccalieu Tickle in memory of John Cabot and his crew. I, along with my wife and some other people, witnessed the event from another ship at Grates Cove that day after which we escorted the *Matthew* through Baccalieu Tickle and along the north shore of Conception Bay to Harbour Grace.

It is unlikely that we will ever know for certain what happened to Cabot's expedition of 1498. Some documents may be found in some archives in the future to enlighten us. For now, we only have the two theories which have been created or concocted by historians over the years. These theories have come about by a study and interpretation of documents and maps in England and Spain. But really we have only scraps of information, some of which make a lot of sense and provide logical explanations, but a logical explanation is not proof, and therefore there is really no way to know for certain what happened to John Cabot's expedition of 1498.

In the following pages we will examine in more detail the information we do have about Cabot's voyage of 1498 and hope that our efforts will throw more light on this important subject. In Williamson's view, "The subject cannot be fairly shirked, since the first navigators between England and North America were the beginners of the United States and the British Commonwealth." [28]

In the first section of his book, *Decades* produced in 1516, the Italian historian, Peter Martyr working in Spain makes no mention of John Cabot although he does give a description of Sebastian Cabot's north-western voyage of 1508-1509. Hence John Cabot, the true discoverer of North America, who made two voyages; one in 1497 and the other in 1498, was not only lost in the literal sense, he was also lost to history. Three and a half centuries would pass before this great Italian navigator and explorer would be truly credited and honoured for the voyages of discovery and exploration now associated with his name.

Notes to Chapter 1

The·Voyage of 1498: An Overview

1. Tryggvi J. Oleson and William L. Morton, "The Northern Approaches to Canada," *The Dictionary of Canadian Biography*, Volume 1 (1000 to 1700), (Toronto, 1966), p. 19.
2. *Loc. cit.*
3. Raleigh A. Skelton, *The Dictionary of Canadian Biography*, Volume 1, p. 150.
4. *Loc. cit.*
5. Pedro de Ayala's Letter. See Document, Number 8.
6. Polydore Vergil on John Cabot. See Document, Number 5.
7. James A. Williamson, *The Cabot Voyages and Bristol Discovery under Henry VII* (Cambridge, 1962), p. 105.
8. Raleigh A. Skelton, *op. cit.*, p. 151.
9. From the patent granted to Alonso de Hojeda. See Document, Number 9.
10. Pedro de Ayala's Letter, *op. cit.*
11. Ian Wilson, *The Columbus Myth* (London, 1991), p. 137.
12. *Loc. cit.* See also Williamson, *op. cit.*, p. 111 and Martin Fernndez de Navarrete, *Colleccion de los viajes y descubrimientos que hieieron por mar los Espanoles*, Volume III (Madrid, 1829), p. 41.
13. Williamson, *op. cit.*, p. 111.

14. *Ibid.*, p. 111-112.
15. Pedro de Ayala's Letter, *op. cit.*
16. See *The Canadian Encyclopedia*, second edition (Edmonton, 1988), p. 312.
17. Williamson, *op. cit.*, p. 113.
18. *Ibid.*, p. 140.
19. *Ibid.*, see Preface, p. vii.
20. Raleigh A. Skelton, *The Cartography of the Voyages* in Williamson, *op. cit.*, pp. 299-305.
21. Ian Wilson, *op. cit.*, p. 143-144.
22. Samuel E. Morison, *The European Discovery of America: The Northern Voyages, 500-1600 A.D.* (New York, 1971), p. 193.
23. *Ibid.*, p. 194.
24. Samuel E. Morison, *The European Discovery of America: The Southern Voyages, 1492-1616* (New York, 1974), p. 206.
25. Wilson, *op. cit.*, p. 145.
26. John Parsons, *Away Beyond the Virgin Rocks: A Tribute to John Cabot* (St. Johns, 1997), See Chapter 3. See also Mike Flynn, "Cabot Died Here," *The Evening Telegram*, St. Johns, February 8, 1995, p. 1.
27. Arthur Davies, "The Last Voyage of John Cabot and the Rock at Grates Cove," *Nature*, Volume 176, November 26, 1955, pp. 996-999.
28. Williamson, *op. cit.*, p. 172.

Chapter 2

John Cabot Sails into History

On August 6, 1497 John Cabot returned to Bristol from his first voyage, and on August 10 he was in London informing King Henry VII of his discoveries in the western ocean. Both Cabot and the English monarch were enthusiastic about the new discoveries and it is apparent that almost immediately planning began for a more ambitious voyage the following spring. From all reports it would appear that the fishermen of Bristol were impressed by the large schools of codfish they had found across the ocean. Howbeit, Cabot himself was more interested in jewels and the spice trade. [1]

In early December 1497 Cabot presented to Henry VII his proposals for a second voyage. All indications were that Cabot on that voyage intended to follow the coastline to the south and west from his 1497 landfall until he came to East Asia, the island of Cipango in the equinoctial region, the source according to Raimondo de Soncino's letter of "all the spices . . . as well as the jewels" in the world. [2] Raleigh A. Skelton says, "This terminology plainly echoes Marco Polo's description of Cipango and the eastern archipelago; and the association of ideas is strengthened by the wording of [John] Day's letter, written about the same time, with which he sent to his correspondent, 'the other book of Marco Polo and the copy of the land which has been found.' " [3] The Spanish ambassador in England, Pedro de Ayala, reporting to the Spanish monarchs on the voyage of 1497 wrote, "What they have discovered or are in search of is possessed by Your Highnesses because it is at the cape which fell to Your Highnesses by the convention with Portugal." [4] What they "are in search of," of course, is a reference to Cabot's voyage planned for 1497, and the "convention with Portugal" is a reference to the Treaty of Tordesillas June 7, 1494 in which the pope divided the Atlantic into Portuguese and Spanish spheres of influence. Skelton believed that "the cape which fell to Your Highnesses" is clearly a reference to Cuba, understood at that time by Columbus to be a cape or large peninsula of Cathay. [5] Cabot's expressed intention then was to find the spices and jewels of East Asia — known by the people of Cabot's time as Cipango and Cathay. Hence, Cabot for all intents and purposes was not interested in fish; he was interested in something much more valuable and his knowledge of the writings of Marco Polo (1254-1324) convinced him that the spices and jewels were there for the taking.

Cabot's first voyage in 1497 also convinced him that he was on the right track towards reaching his goal which was Cathay. As James A. Williamson wrote in his classic study of Cabot, "In the winter of 1497-98, the interval between the voyages, John Cabot was quite certain that Asia faced him across the Atlantic." [6]

John Day's letter is probably the most significant document in existence regarding Cabot's first voyage in 1497. Another letter which can be found in the same archives in Spain dated July 25, 1498 and written by Pedro de Ayala is undoubtedly one of the most important documents pertaining to Cabot's second voyage of 1498. Ayala says that Cabot left in May 1498 and he infers that it was Cabot's intention to return by September 1498. Ayala wrote his letter to the Spanish monarchs, Ferdinand and Isabella, in late July while Cabot was either dead or somewhere on the western side of the Atlantic. [7]

Ayala at the time was walking a kind of tightrope. As Spanish ambassador he had to keep on good terms with Henry VII of England and at the same time he had to keep his monarchs informed of everything going on in England particularly with regard to discoveries and explorations in the western ocean. Ian Wilson says that, at the same time, Ayala had another important and delicate job to do: "Currently Ayala was carefully preparing the ground for a delicate treaty of alliance between England and Spain, to be marked by the marriage of Ferdinand and Isabella's daughter, Catherine, to Henry VII's eldest son and heir, the prince Arthur, brother to the future Henry VIII." [8]

But in his letter of July 25, 1498, it is apparent that Ayala was mainly concerned with Cabot's new expedition of 1498 and the effect it would have on discoveries already made by Spain. In fact the tone in Ayala's letter is somewhat bitter. Ayala informs his monarchs that the king of England has often spoken to him of Cabot's voyage and the land discovered in 1497, but says Ayala, "I told him that I

believed the islands were those found by Your Highnesses, and
although I gave him the main reason, he would not have it." [9] Williamson believed that, "Ayala was highly indignant at the whole
geographical argument, which he saw as an attack upon the achievement of Spain." [12] Apparently Ayala saw his letter being of great importance and significance as well as of current relevant interest.
Hence he goes into great detail in an attempt to inform his
monarchs:

> I think Your Highnesses have already heard how the king of
> England has equipped a fleet to explore islands or mainland which
> he has been assured certain persons who set out last year from
> Bristol in search of the same have discovered. I have seen the map
> made by the discoverer, who is another Genoese like Columbus,
> who has been in Seville and at Lisbon seeking to obtain persons to
> aid him in this discovery . . . Having seen the course they are steer-
> ing and the length of the voyage, I find that what they have dis-
> covered or are in search of is possessed by Your Highnesses
> because it is at the cape which fell to Your Highness by the conven-
> tion with Portugal. [11]

Ayala seemed to be well informed and knew that Cabot was of
Genoese birth. We also learn that Cabot had approached other
people in Seville, Spain and Lisbon, Portugal for support for his
planned expedition of 1498. No doubt this had been done in secret,
and when Cabot could not get the support he wanted he again turned
towards England and King Henry VII. So Ayala was well informed
about Cabot's voyage of 1498 and a few months after Cabot left he so
informed his Spanish monarchs.

An important question in the quotation above from Ayala's letter
is related to the reference to "the cape which fell to Your Highnesses." As mentioned earlier some historians think that the cape
referred to was Cuba which Columbus up to this time had not discovered (or did not want to report) was an island. Apparently
Columbus objective was to lead the Spanish monarchs to believe that
Cuba was part of the Asiatic main continent. [12] Hence it was Ayala's
opinion — and he was correct — that Cabot on his expedition of
1498 was clearly heading south and west to the tropical lands already
in possession of Spain. Ayala in July 1498 was fully cognizant of the
fact that Cabot's intentions could create a rather dangerous situation.
Hence it is clearly implied in Ayala's letter that Cabot would end up
in the Caribbean Sea, and very likely would clash with the Spaniards.

When Cabot's expedition left Bristol in May 1498, he assumed that Asia and the riches he was seeking there lay to the south and west of the land he had discovered in 1497. His mind was made up and his voyage of reconnaissance in 1497 had convinced him that the land of the Great Khan and the riches of Cipango were within his reach. [13] Ayala's letter says that Cabot's expedition encountered a storm and one of the ships was so badly damaged that it had to return to an Irish port presumably for repairs, but says Ayala in reference to Cabot: "The Genoese kept on his way." The damaged ship may or may not have continued the voyage across the Atlantic but, according to Ayala, Cabot must have been on one of the other four ships and hence continued the journey. [14] The historian, Polydore Vergil, says in his *Anglica Historia* (1513): "John set out in this same year and sailed first to Ireland. Then he set sail towards the west." [15] This may or may not imply that Cabot was on the ship that was damaged in "a great storm" and was forced to seek shelter in an Irish port. Since there is no English record of any ship from Cabot's 1498 expedition returning to Bristol, it has to be assumed that the damaged ship presumably after repairs resumed the voyage towards the west hoping, one would assume, to eventually rendezvous with the other four ships. Williamson sums up this matter very nicely in these words:

> After this evidence of the ship returning to Ireland there is no direct information of the proceedings of any part of the expedition. The chronicles that mention its start merely say that no news had been received by mid-September . . . Perhaps no further tidings ever did come, perhaps the ships returned to Bristol, certainly they brought no rich cargoes of spices. The chroniclers were writing only of outstanding London events. If there was a straggling back to Bristol with a report of commercial failure, a London writer may not have thought it worth while to make a new reference to the subject . . . The silence of the chronicles proves neither that the whole expedition was lost nor that part of it returned. [16]

Williamson believed that it was unlikely that all five ships were lost at sea. In his opinion, "There is no instance of a multi-ship expedition having been entirely wiped out by an unknown disaster; and we are entitled to say that the odds were heavily against it in 1498." [17]

If we assume that all five ships made it across the Atlantic, then the big question we have to wrestle with is this: what happened to them? And what was John Cabot's ultimate earthly destiny? Obviously, one or more, even all five, could have been wrecked on an unknown shore. One or more could have pushed on south and west in an effort to fulfil Cabot's original intention to reach the land of spices and jewels which, after all, was the main objective of the 1498 expedition.

The expedition had enough supplies to last for a year and on board there were considerable trading goods as well. Wilson says, "The very character of this merchandise indicates that it was intended for trade with the 'Asians' with whom this time Cabot anticipated proper contact, and that spices, not fish, were the commodity with which they hoped to return." [18] Furthermore, it was Cabot's intention to have the expedition return with its riches not to Bristol but to London. This is stated clearly in Soncino's letter of December 18, 1497: "Before very long they say that his Majesty will equip some ships, and in addition he will give them all the malefactors, and they will go to that country and form a colony. By means of this they hope to make London a more important mart for spices than Alexandria. The leading men in this enterprise are from Bristol, and great seamen, and now they know where to go . . ." [19]

If one or more of Cabot's ships in 1498 had reached North America in the vicinity of the Bay of Fundy, where they had made landfall in 1497, then their intention was clearly to sail south and west into the area of Spanish influence in their attempt to find the riches of East Asia. Again this is quite evident from Soncino's letter of December 14, 1497 to the Duke of Milan:

> But Messer Zoane (meaning John Cabot) has his mind set upon even greater things, because he proposes to keep along the coast from the place at which he touched, more and more towards the east, until he reaches an island which he calls Cipango, situated in the equinoctial region, where he believes that all the spices of the world have their origin, as well as the jewels. [20]

When Soncino in the above quotation says "towards the east," he was referring to East Asia which would to him be "the east", in reality though from our vantage point he is talking about Cabot's expedition moving westward. This might sound rather confusing even though over the years several historians have attempted to clarify this matter. In another telling phrase Soncino with reference to Cabot and his men says, "Now they know where to go." The truth is, of course, they only thought they knew where to go, and their going eventually led one or more of the ships into the waters of the Caribbean Sea and in contact with the Spaniards. All this clearly indicates that Cabot, whether intentionally or unintentionally, was heading into the area of Spanish influence towards the lands that Colombus had discovered and which the Spaniards regarded as rightfully and legally theirs according to the papal bull as set forth in the sacred Treaty of Tordesillas of 1494. Many historians are now convinced that Cabot in his quest for spices and jewels entered an unknown and a dangerous territory and by so doing sealed his fate and that of those who accompanied him. [21]

Of significance in the story of Cabot's expedition of 1498 are the voyages by the Portuguese explorer, Gaspar Corte Real. In 1500 this explorer landed on the coast of North America in about latitude 50^0 N. Corte Real's first voyage like Cabot's of 1497 was apparently one of reconnaissance for the next year, 1501, this explorer set out from Portugal as head of a three-ship expedition. Something happened on this voyage of 1501 which may be of great significance as regards Cabot's voyages of 1497 and 1498, but most likely of 1498. On October 19, 1501 the Venetian ambassador to Portugal, Pietro Pasqualigo, a brother to Lorenzo Pasqualigo who had written a letter to his brothers about Cabot's first voyage of 1497, sent to his officials in Venice a rather detailed report about Corte Real's expedition of 1501. The following are four extracts from Pasqualigo's report:

> On the eighth of the present month arrived here one of the two caravels which this most August monarch sent out in the year past under Captain Gaspar Corterat [i.e. Corte Real] to discover land towards the north; and they report that they have found land two thousands miles from here, between the north and the west, which never before was known to anyone. They examined the coast of the same for perhaps six hundred or seven hundred miles and never found the end, which leads them to think it mainland. This continues to another land which was discovered last year in the north. The caravels were not able to arrive there on account of the sea being frozen and the great quantity of snow. They are led to the same opinion [i.e. that the land was mainland] from the considerable number of very large rivers which they found there, for certainly no island could ever have so many nor such large ones . . . They have brought back here seven natives, men and women and children, and in the other caravel, which is expected to be coming from hour to hour are coming fifty others. They resemble gypsies . . . are clothed in the skins of various animals, but chiefly otters . . . They speak, but are not understood by anyone, though I believe that they have been spoken to in every possible language. In their land there is no iron, but they make knives out of stones and in like manner the points of their arrows. And yet these men [i.e. the Corte Real expedition] have brought from there a piece of broken gilt sword, which certainly seems to have been made in Italy. One of

the boys was wearing in his ears two silver rings which without
doubt seem to have been made in Venice, which makes me think it
to be mainland, because it is not likely that ships would have gone
there without their having been heard of. They have great quantity
of salmon, herring, cod and similar fish. They have also great store
of wood and above all of pines for making masts and yards of ships
. . . [22]

It is evident from this report that the "mainland" visited was likely
the coast of Nova Scotia or New England. Of particular significance
is the reference to the "very large rivers." There are several large
rivers in the Bay of Fundy region. The piece of the broken sword and
the earrings very likely came from Cabot's expedition of 1498. It will
be recalled that Cabot's second expedition carried with it a large
quantity of merchandise for trading purposes. From this it can be
deduced that at least one of Cabot's ships in 1498 reached the North
American coast and made landfall.

It is very unlikely that the two items in question came from
Cabot's first voyage. On that voyage he made landfall but did not
come in contact with the natives, and it is even more unlikely that the
piece of sword and earrings came from the Bristol fishermen
engaged in their fishing operations near the Avalon Peninsula their
"Island of Brasil." The question still remains as to how the natives
acquired these items from Cabot's expedition in 1498. Did they
engage the Englishmen in a fight and take the items or did the natives trade them with Cabot and his men in return for furs or some
other item? Wilson comments on the matter this way:

> The only viable alternative is that the items came from Cabot's
> 1498 expedition — i.e. that one or more ships did survive the storm
> and reach the other side. Yet even this possibility raises more ques-
> tions than it answers.
>
> For instance, did the surviving ships land up on the shores of
> Nova Scotia, only for the crewmen then to be murdered by the na-
> tives, and their belongings plundered? Could the sword perhaps
> even have been Cabot's own, broken during a last desperate skir-
> mish. But if so, why should the same natives have shown no similar
> hostility towards the Portuguese on their arrival?
>
> Alternatively, might Cabot's expedition have arrived peacably
> and in good order, and simply used the sword hilt and rings — the
> very sort of trifles with which the Londoners reportedly stocked
> Cabot's ships — to barter for food or furs before making their way
> further south? On its own, without any more evidence to back it
> up, this latter possibility might seem still very tenuous. [23]

Williamson believed that the natives acquired the goods in a trading transaction, and furthermore, he believed that the sword and earrings came from the Cabot expedition of 1498. Williamson comments as follows: "However, the story is an indication that part of the expedition of 1498 reached the neighbourhood explored in 1497. If the Indians gained the things in a fight, it might mark the end of Cabot and his crew. But the supposition is very speculative. It was more probably by barter." [24]

Effectively we have no clear-cut documented information of the discoveries made by the Cabot expedition of 1498, although there are several documents in two Spanish archives which seem to indicate that at least some of the ships in Cabot's voyage did advance south and ultimately strayed or deliberately moved into a region which the Spanish regarded as their own. Cabot's ships were provisioned for a year, so it is clear that a much longer voyage than that of 1497 was planned. Nothing was heard of the expedition by September 1498 and the people of Bristol were uneasy that something had gone wrong. Nobody knows for certain what happened to the five ships. They may have kept together along the coast of North America, or they may have become separated and some may have been lost, but in all probability at least one ship continued the voyage into the year 1499. Cabot had indeed sailed into history, but the day and date when he crossed the threshold between time and eternity may never be known.

Notes to Chapter 2

John Cabot Sails into History

1. Raleigh A. Skelton, *The Dictionary of Canadian Biography*, Volume 1, p. 150.
2. *Loc. cit.*
3. *Loc. cit.* See also Raimondo de Soncino's Letter. See Document Number 2.
4. *Loc. cit.*
5. *Loc. cit.*
6. James A. Williamson, *The Cabot Voyages and Bristol Discovery under Henry VII*, (Cambridge, 1962), p. 99.
7. Pedro de Ayala's Letter. See Document, Number 8.
8. Ian Wilson, *The Columbus Myth* (London, 1991), p. 118.
9. Pedro de Ayala's Letter, *op. cit.*
10. on, *op. cit.*, p. 89.
11. Pedro de Ayala's Letter, *op. cit.*
12. Skelton, *op.cit.*, p. 150.
13. Williamson, *op. cit.*, p. 99.
14. Pedro de Ayalas Letter, *op. cit.*
15. Polydore Vergil on John Cabot. See Document, Number 5.
16. Williamson, *op. cit.*, pp. 104-105.
17. *Ibid.*, p. 105.
18. Wilson, *op. cit.*, pp. 121-122.
19. Raimondo de Soncino's Letter. See Document, Number 2. Alexandria was the great trading centre of Egypt.
20. *Loc. cit.*
21. Wilson, *op. cit.*, p. 137.
22. Pietro Pasqualigo's Report. See Document, Number 10.
23. Ian Wilson, *John Cabot and the Matthew* (Bristol and St. John's, 1996), p. 48.
24. Williamson, *op. cit.*, p. 107.

Chapter 3

Englishmen in South America

There are three or four possibilities as to the ultimate fate of John Cabot and the crews that made up his expedition in 1498. One possibility is that, with the exception of the ship that returned to an Irish port, they were all swallowed up by a vicious storm in the North Atlantic. They could have made it across the Atlantic and were shipwrecked on an unknown shore. They could have made contact with the natives and were all killed in a fight over their trade goods, or they could have gone south and were all murdered by the Spaniards. There is a theory that at least one ship, possibly the one Cabot was on, was lost in Baccalieu Tickle near Grates Cove in Newfoundland. [1]

The American historian, Samuel E. Morison, believed the expedition was wiped out in a storm. English historians, James A. Williamson and Raleigh A. Skelton, lean towards the theory that the expedition, or at least part of it, went too far south and was eliminated by the Spaniards. Modern writers, such as Ian Wilson and Peter Firstbrook, seem to be inclined in that direction as well and Arthur Davies, an English geographer, opts for the theory that Cabot ended his days in Baccalieu Tickle on the east coast of Newfoundland.

Peter Firstbrook in his recent book on Cabot makes a statement and then poses three questions:

> Cabot sailed from Bristol for the last time in early May 1498, but what happened to him and his fleet of five ships is still a mystery. Did he return quietly to England having failed, a broken and destitute man? Or did he and his ships perish during the crossing leaving no trace? Or is there another explanation: did Cabot and some of his ships succeed in crossing the North Atlantic a second time, only to die by other, more sinister, means? [2]

28

As we have seen from Pedro de Ayala's letter, the Spaniards did not look too favourably on Cabot's voyages in the western ocean and the belief that Cabot might have wandered into Spanish territory certainly had the makings of a dangerous situation. The Spanish were ready and waiting for any intruders who might come their way, and there was one Spanish adventurer in particular that may have had much to do with Cabot's ultimate fate. The Canadian historian, Brian Cuthbertson, in his recent book makes the following reference to that piratical Spaniard: "A 1499 Spanish expedition under the command of Alonso de Hojeda may have encountered English ships in the Caribbean. These could only have been part of Cabot's expedition. De Hojeda spent the voyage 'killing, robbing and fighting' so the fate of any English he came across can be readily imagined." [3]

Samuel E. Morison in his book, *The European Discovery of America: The Southern Voyages 1492-1616* (1974) gives a long detailed description of Hojeda, the dashing young "golden boy" adventurer who, at the age of twenty-seven, was the terror of the Caribbean. He not only murdered the natives of the area but his own people as well. Cuthbertson's reference to "killing, robbing and fighting" comes from Morison who gives the impression that Hojeda was a rapist and torturer as well. [4]

Spanish documents of the period indicate that Hojeda's objective was two-fold: to assist Columbus in the Caribbean to administer the Spanish territory, and to deal effectively with any Englishmen who may have wandered into the Spanish area. [5] On his 1499 expedition, Hojeda landed on the coast of South America east and south of what is today Trinidad and then with his two ships worked his way north and west past Trinidad, the island of Margarita, the present day islands of Bonaire and Curacao and into the large gulf which the natives referred to as Coquibacoa. [6] He named the area Venezuela, or Little Venice because there was a native village nearby built on stilts over the water. For their amusement Hojeda and his crew murdered a few natives, looted their houses, and forcibly took a young woman to serve as his mistress. [7]

It is believed that in this vicinity this Spanish buccaneer encountered some English adventurers either in one of the harbours near Coquibacoa or sailing eastward along the coast. Hojeda's assignment from his Spanish monarchs on his 1499 expedition was to locate any Englishmen in the area and to deal with them in the appropriate manner. At this point some questions that beg for answers are: did Hojeda encounter any Englishmen along the Spanish Main and what really happened to the Englishmen?   Where is the evidence, if any, that gives us some insight into the details of Hojeda's voyage in the summer of 1499?   Common sense and the implications in Spanish documents tell us that if indeed Hojeda encountered Englishmen in South America, then they had to be part of Cabot's expedition of 1498 simply because, with the possible exception of some Bristol fishermen in the Avalon Peninsula area of Newfoundland (The Island of Brasil), there were no other Englishmen anywhere on the coast of North or South America at the time. Hence some of Cabot's ships had sailed south the full length of the coast of North America and across the Caribbean to the South American coast.

The English historian, James A. Williamson believed that Hojeda's coming in contact with Englishmen on the northern coast of South America is supported by two important pieces of information. It is supported by the reference in a book written in 1829 by the Spanish historian Martin Fernndez de Navarrete, and it is further supported by a patent or licence granted to Hojeda by the Spanish monarchs on June 8, 1501. [8]   Although undocumented, the reference in the writings of Navarrete certainly sounds convincing, and why he did not name the source for his statement is in the words of Ian Wilson, "one of the most tantalisingly enigmatic in the whole history of America's discovery." [9]   In 1962 Williamson wrote, "One thing may be added, that the Spanish information need not have been obtained through England. If Navarrete is right, it could have been gathered from the Englishmen encountered at Coquibacoa, and these men may never have reached home." [10]   As for Hojeda's patent in 1501, it is clear that the Spanish monarchs anticipated that more Englishmen might be encountered in the Caribbean. The wording is very clear: "So that you may stop the exploration of the English in that direction." But by June 1501 when this patent was granted there were no Englishmen in the area. Hojeda had eliminated them on his voyage of 1499 for this too is clearly implied in Hojeda's licence or patent. [11]

All of this adds up to a sad story for, if Cabot and his men had made it all the way from Bristol to South America and along the coast of North America, imagine the geographical information they would have recorded; for if indeed they really did make it to South America, then these Englishmen would have been the first to explore the coast of North America, the Spaniards up to that time not having gone north of islands of the Caribbean or west of Coquibacoa or Cape de la Vela. The following extracts from one of Wilson's books puts the whole matter in proper perspective:

> So had Cabot and/or those who had survived from his 1498 voyage, diligently searching, like Columbus, for Marco Polo's 'Cipango . . . in the equinoctial region,' been all this time making their way steadily southwards down the eastern American coastline, only to come face to face with Hojeda and his band of Spanish desperados in Coquibacoa?
> It is a tantalising thought that had they indeed done so they would have been virtually bound to have acquired more knowledge of America's geography than any other Europeans at this time, their peregrinations southward inevitably convincing them that this had to be a whole new continent . . . Hojeda's well-practised ruthlessness towards natives and even fellow-Spaniards is indication enough that he and his men would have had little compunction about liquidating any such stray group of Englishmen, particularly since he could justify his actions on the grounds that they were trespassing on territory already allotted to Spain. [12]

The big question then that begs an answer is what happened to the geographical knowledge gained by Cabot and his expedition of 1498 or at least the remnants of that expedition? There is no evidence that any ship from Cabot's expedition made it back to England to relay their findings to king Henry VII. My interpretation is that the knowledge gained by the certain Englishmen who encountered Hojeda's expedition in 1499 was later incorporated in La Cosa's famous map dated 1500. How else could La Cosa's map show such accuracy regarding places he himself or his fellow Spaniards had never been? Juan de la Cosa had been a member of Hojeda's expedition of 1499.

When Hojeda returned to Spain in the summer of 1500, instead of being executed or at least severely reprimanded for his many acts of piracy, he was given a warm welcome, and the following summer was given a new patent or licence authorizing him to return to Coquibacoa and establish a proper trading post. He was also officially given the title of "Governor of the Province of Coquibacoa." [13]

Hence on June 8, 1501, the Spanish monarchs issued a new licence to Hojeda to "pursue his discoveries" in the Caribbean Sea, an area that Spain still wished to explore and which they claimed exclusively. The wording of Hojeda's new licence gives every indication that on his earlier voyage in 1499 he had indeed encountered certain English explorers. Hojeda's licence which is dated and clearly explicit mentions "the English" three times. [14] It would appear that Ferdinand and Isabella were rather obsessed with the English intruders in their territory.

Like John Day's letter and Pedro de Ayala's letter, the original copy of the patent dated June 8, 1501 and granted to Hojeda can be found in the Simancas archives in Spain. Along with Ayala's letter this document may be one of the most significant documents in existence relating to Cabot's expedition of 1498. It is significant in that Hojeda is given explicit orders to "stop the exploration of the English" in the Caribbean region. Hojeda was instructed as follows:

> Item: that you go and follow that coast which you have discovered, which runs east and west, as it appears, because it goes towards the region where it has been learned that the English were making discoveries; and that you go setting up marks with the arms of their Majesties, or with other signs that may be known, such as shall seem good to you, in order that it be known that you have discovered that land, so that you may stop the exploration of the English in that direction. [15]

This information is significant and is support for the Spanish historian, Navarrete, who in 1829 claimed that Hojeda encountered Englishmen on his voyage of 1499. Another statement in Hojeda's licence indicates that "green stones" or emeralds had been found in the region west of Coquibacoa, a region which the Spaniards up to that time had not explored. So how did the Spaniards obtain the precious stones? The implication is that the stones were taken from members of Cabot's expedition who had explored the region west of Coquibacoa and had unfortunately encountered Hojeda as they sailed towards the east. And yet another item in the licence refers to Englishmen in the area; in fact, Hojeda is given a grant of land for services rendered "for the stopping of the English."[16] Wilson says that, "there seems . . . more than a touch of adroitness in the way the Spanish sovereigns concluded the licence by rewarding Hojeda, ostensibly not for what he had done, but for what he was to do." [17] The Spanish monarchs wished to show their gratitude to Hojeda by giving him a large grant of land on the island of Hispaniola stated in the licence in such a way as if it were for future services whereas in reality it was very likely for services already rendered:

> Likewise their Majesties make you a gift in the island of Hispaniola [present day Haiti and the Dominican Republic] of six leagues of land with its boundary, in the southern district which is called Maquana, that you may cultivate it and improve it, for what

you shall discover on the coast of the mainland for the stopping of the English, and the said six leagues of land shall be yours forever.[18]

The items found by the Corte-Real expedition in 1501, the Spanish historian's reference in 1829 to Englishmen in the Caribbean area, and the rather explicit information in Spanish documents particularly Hojeda's licence of 1501 certainly present a strong argument that at least part of Cabot's expedition of 1498 not only reached North America but South America as well. This evidence seems to be very convincing, and presents us with a logical explanation in our attempt to solve the problem of John Cabot's expedition of 1498. In addition, there is also the evidence that can be extrapolated from another important document, the La Cosa map of 1500, produced by a man who sailed on one and possibly two of Hojeda's expeditions. [19] Still despite all this evidence or what most certainly appears to be evidence or can be interpreted as such, Samuel E. Morison, that most cantankerous of American historians, will have none of it, and in 1974 dismissed all of the foregoing in his own inimitable way:

> A minor object of this voyage [i.e. Hojeda's third expedition of 1502] was to root out any Englishmen who might be found on the Spanish Main; and as nothing had been heard from John Cabot's 1498 voyage, it was feared that he might have tried to start a colony in the Caribbean. But no Spanish explorer of this area ever saw hide or hair of any Englishmen. [20]

Notes to Chapter 3

Englishmen in South America

1. Arthur Davies, "The Last Voyage of John Cabot and the Rock at Grates Cove," *Nature*, Volume 176, November 26, 1955, pp. 996-999.

2. Peter Firstbrook, *The Voyage of the Matthew: John Cabot and the Discovery of North America* (Toronto, 1997), p. 131.

3. Brian Cuthbertson, *John Cabot and the Voyage of the Matthew* (Halifax, 1997), p. 50.

4. Samuel E. Morison, *The European Discovery of America: The Southern Voyages, 1492-1616* (New York, 1974), p. 187.

5. Patent granted by the Spanish sovereigns to Alonso de Hojeda, June 8, 1501. See Document, Number 9.

6. Morison, *op. cit.*, p. 189.

7. *Loc. cit.*

8. James A. Williamson, *The Cabot Voyages and Bristol Discovery under Henry VII* (Cambridge, 1962), p. 111. See also Document, Number 9.

9. Ian Wilson, *The Columbus Myth*, (London, 1991), p. 137.

10. Williamson, *op. cit.*, p. 111-112.

11. See Document, Number 9.

12. Wilson, *op. cit.* p. 137.

13. See Document, Number 9.

14. *Loc. cit.*

15. See Document, Number 9.

16. *Ibid.*

17. Wilson, *op. cit.*, p. 139.

18. See Document, Number 9.

19. Firstbrook, *op. cit.* p. 140-141.

20. Morison, *op. cit.*, p. 191.

Chapter 4

Juan de la Cosa's Enigmatic Map

Both John Day in his letter to Columbus and Pedro de Ayala in his letter to Ferdinand and Isabella of Spain mention maps and a globe that John Cabot had made relating to his voyage of 1497.[1] Raimondo de Soncino in his letter to the Duke of Milan dated December 18, 1497 reported that, "This Messer Zoane (meaning John Cabot) has the description of the world in a map, and also in a solid sphere, which he has made, and shows where he has been." [2] As far as maps depicting the voyage of 1497 were concerned Cabot had indeed done his homework. There is no reason to believe that the same was not done as regards the voyage of 1498. In fact, on that voyage Cabot was probably more meticulous as regards cartography than he had been on the voyage the year before. Hence there are three questions that we must wrestle with. What happened to the map and sphere Cabot made after his first voyage? Assuming he made a map of his second voyage, did it survive and what happened to it? All this begs another question. Could all these cartographical efforts by Cabot and his men have ended up in the hands of some Spanish cartographer who might have used them to create his own map? It is quite possible that Juan de la Cosa's famous map of 1500 had its origin in the cartographical efforts of John Cabot.

In 1832 an aristocratic German, Baron Walckenaer, found an old map in a curio shop in Paris. The map, drawn on oxhide and measuring about three feet by six feet, bore the following inscription: *Juan de la Cosa la fizo en el Puerto de S. Ma. en ano de 1500* ("Juan de la Cosa made this at Puerta de Santa Maria in the year 1500").[3] Walckenaer was wise enough to realize that he had come across a historical and geographical treasure, a document which depicted even if roughly, the coastline of North and South America. What the Baron might not have known is that the La Cosa map is one of the

35

most significant maps in existence regarding a study of John Cabot's voyages in 1497 and 1498.

After Baron Walckenaer's death in 1853 the La Cosa map was purchased from his estate by the government of Spain and is now housed in the Spanish Naval Museum in Madrid. Juan de la Cosa who created the map had sailed on Columbus' voyage of 1493 and served as a cartographer on Hojeda's voyage in 1499. It was the voyage commanded by Hojeda which might have encountered some of the ships of Cabot's ill-fated voyage of 1498.

George E. Nunn pointed out in 1934 that there is a problem with the La Cosa map in that it is made up of two distinct sections — the Old World and the New World — with a meridian running through the Atlantic Ocean. Each section is drawn to a different scale. Because the scale is not uniform, it is therefore virtually impossible to adequately relate the latitudes on both sides of the Atlantic or the distances on the map with those reported in the primary sources to have been covered by Cabot on his first voyage.[4] Raleigh A. Skelton was of the opinion that the La Cosa map very likely reflected information from Cabot's two voyages.[5]

Samuel E. Morison believed that the La Cosa map could be made to fit any area on the east seaboard of North America and chided historians for their efforts in this regard: "Theorists treat it like rubber — they squeeze, stretch, twist, and telescope it to fit anything from a hundred to a thousand miles and even turn it sideways or upside down." [6] Most historians accept the year 1500 as the authentic date for this map. That being so, then La Cosa made his map the next year after his voyage with Hojeda in 1499. It was on this voyage that contact was likely made with all or part of the Cabot expedition of 1498 and it was from these Englishmen that La Cosa got sufficient information to enable him to do his map with a certain degree of accuracy, something he might not have been able to do had he not come into contact with the Cabot expedition. What can be extrapolated from all this is that the La Cosa map is made up of a combination of geographical and cartographical knowledge gained from both the English and the Spanish. [7] There is no way that La Cosa could have drawn the map of North America as he did it without input from both Cabot expeditions.

Furthermore, James A. Williamson pointed out in 1962 that La Cosa's depiction of the coastline in South America from Coquibacoa around the Gulf of Darien is, "Remarkably good," its only fault being, "that it does not make the Gulf of Darien extend deeply enough to the southward."[8] Ian Wilson says this section of the coastline is "one of the most uncannily accurate of the whole map."[9] All this is truly remarkable when it is realized that La Cosa was never in this area. In his book in 1991 Wilson asks the obvious questions: "So how could La Cosa have obtained his information about the coast west of Coquibacoa, and could it have been from the same source that Hojeda obtained his information about the region's gold, emeralds and pearls?"[10] More significantly the same kind of question can be asked regarding all the coastline of North America north of the Caribbean Sea. We can accept the fact that some of La Cosa's information could have come from Cabot's voyage of 1497 by way of reports and maps sent from England, but the information that La Cosa had about the rest of North America from Maine to Florida, and the area west of Coquibacoa must have come from Cabot's expedition of 1498.[11]

The implication here is that not only did the Cabot expedition of 1498 explore the east coast of North America, they very likely explored the area west of Cape de la Vela possibly even as far west of what is now Panama.[12] This being the case then, it can be said that Cabot and his men (if indeed Cabot was on a ship that made it into the Caribbean) had considerable input into the world famous and enigmatic map drawn by Juan de la Cosa in 1500. How else could La Cosa have known about regions which he and his fellow Spaniards had never visited? How else could he have labelled a large area off the coast of North America as "mar descubierto por inglese" ("sea discovered by the English"), if he had not had contact with Englishmen and made use of their geographical and cartographical information? Wilson puts the matter this way:

> Historians mostly agree that La Cosa must somehow have received some surprisingly authoritative information of the discoveries made by Cabot, but the really thorny question is whether this derived merely from Cabot's voyage of 1497, or whether the map embodies unique data from Cabot's altogether more mysterious voyage of 1498 . . . The key question is whether such features as marked on La Cosa's map could have derived from discoveries made by John Cabot and/or the survivors of his 1498 expedition. Could these survivors have carefully charted everything they found on working their way south (inevitably planning to bring news of it all back to England), only for it all to fall into La Cosa's hands as a result of their fatal encounter at Coquibacoa?[13]

If we accept the premise that La Cosa got his information from

Cabot or members of his crew, then the other big question the answer to which has not been adequately established is whether the information came from Cabot's voyage of 1497 by way of Pedro de Ayala, the Spanish ambassador in London at the time, or whether it came directly from the Cabot expedition of 1498. One thing is certain, and that is that La Cosa did not have much time to work on his map. He was on Hojeda's expedition until late 1499, and it was the next summer at Puerto de Santa Maria in Spain that he drafted and labelled his map.

If Hojeda in 1499 had eliminated the Englishmen he encountered in the Caribbean and then took their charts or maps, then surely this information would have been common knowledge in Spain as soon as the Hojeda expedition returned. But this was not so. The other side to this matter in 1500 was that Spain did not want any information in circulation about any Englishmen encountered and eliminated in the Caribbean.[14] Very likely it was the intention of Ferdinand and Isabella to make certain that no information was leaked because of the attempted alliance with England against the common enemy France. That alliance was to be cemented by the marriage of the Spanish princess, Catherine, and the English prince, Arthur. In Wilson's words, "It would have been absolutely imperative that news of any murder of Englishmen by Spaniards, however well-justified, should not be allowed to leak to the outside world."[15]

There is another interesting twist to the story of the La Cosa map. Williamson points out that the map indicates clearly that with regard to English and Spanish areas of discovery there is a noticeable falsification of the latitudes.[16] The latitudes between the English area of North America and the Spanish area of the Caribbean Sea is, according to Williamson, out by approximately twelve degrees. Hence there was by La Cosa a deliberate "cooking of the latitudes" to deceive somebody, or in Wilson's words, "To persuade somebody that any English movement southwards was seriously out of order." [17] Therefore any Englishmen moving southward would appear to be, "that much more of a flagrant trespass upon Spanish territory." [18] Williamson considered La Cosa to be not only a good cartographer for his time but also a good pilot, and in this historian's opinion the matter of the falsified latitudes was very likely well-calculated and deliberate. In this historian's judgement:

> La Cosa was a pilot of repute who had been three times to the West Indies, including twice with Columbus. He could not have made such an error in good faith.[19]

Some historians believe that La Cosa's attempt to deceive regarding the latitudes was an attempt to justify Hojeda's actions in case the information was leaked that the Englishmen in the Caribbean had been eliminated by the Spaniards.[20] In the end no such leak occurred and Catherine of Aragon married Prince Arthur on November 24, 1501. Arthur died the next year and, in time, Catherine became the first wife of the famous or infamous King Henry VIII.

La Cosa's map is of particular significance for historians in that it is the first world map to show, even if in a primitive way, the eastern outline of the North American continent. It is also significant in that it provides additional evidence that at least part of Cabot's expedition of 1498 did indeed reach as far south as what is today the coast of Colombia and Venezuela where their encounter with a Spanish expedition proved fatal. That encounter, while fatal for the Cabot expedition, was a stroke of good luck for the Spanish cartographer, Juan de la Cosa, who learned enough from the Englishmen and their charts to enable him to finish his map with a degree of accuracy that otherwise would have been impossible. It is left to the cartographer, Raleigh A. Skelton, to sum up this matter:

> If any cartographic evidence for the course of John Cabot's ships in 1498 exists, it can be sought only in the coastline drawn by La Cosa, in a general WSW direction, from *Mar descubierta por inglese* to the latitude of Cuba . . . many students have associated this delineation with the voyage of 1498, identifying certain details of it with real geographical features. It may be conceded that, if La Cosa is here recording the results of a voyage, that voyage is much more likely to be John Cabot's of 1498 than any other; but, beyond this, dogmatism is justified. [21]

Notes to Chapter 4

Juan de la Cosa Enigmatic Map

1. John Day's Letter. See Document, Number 3.
2. Pedro de Ayala's Letter. See Document, Number 8.
3. Ian Wilson, *The Columbus Myth*, (London, 1991), p. 141.
4. Nunn, *The mappemonde of Juan de la Cosa: A critical investigation of its data*, (Jenkintown, Pennsylvania, 1934), p. 60-61.
5. Raleigh A. Skelton, "The Cartography of the Voyages," in James A. Williamson, *The Cabot Voyages and Bristol Discovery under Henry VII*, (Cambridge, England, 1962), pp. 298-299.
6. Samuel E. Morison, *The European Discovery of America: The Northern Voyages, 500-1600 A.D.*, (New York, 1971), pp. 238-239.
7. James A. Williamson, *The Cabot Voyages and Bristol Discovery under Henry VII*, (Cambridge, England, 1962), pp. 107-108.
8. *Ibid.*, p. 108.
9. Wilson, *op, cit.*, p. 145.
10. *Ibid.*, p. 145-146.
11. *Ibid.*, p. 147.
12. Williamson, *op. cit.*, p. 108-109.
13. Ian Wilson, *John Cabot and the Matthew*, (Bristol and St. John's, 1996), pp. 53 and 56.
14. *Ibid.*, p. 57.
15. *Loc. cit.*
16. James A. Williamson, "The early falsifications of Western Indian latitudes," *The Geographical Journal*, March 1930.
17. Wilson, *The Columbus Myth, op. cit.*, p. 149.
18. Wilson, *John Cabot and the Matthew, op. cit.*, p. 57.
19. Williamson, *The Geographical Journal*, March 1930.
20. Wilson, *The Columbus Myth, op. cit.*, pp. 148-149. See also Williamson, *The Geographical Journal*, March, 1930.
21. Raleigh A. Skelton, *op. cit.*, pp. 298-299.

Chapter 5

The "Cabot Rock" at Grates Cove

One of the most interesting stories connected with John Cabot's expeditions to the New World has to do with the so-called Cabot Rock at Grates Cove, Trinity Bay, Newfoundland on the northern part of the Avalon Peninsula. The "Cabot Rock" story is a tall tale and I suppose there is not a professional historian alive today who takes it seriously. The local people in and around Grates Cove have always connected the rock with Cabot and some are greatly offended when they encounter people like myself who give little or no credence to it at all.

Francesco Alusio, in a book about Cabot published in 1997 and with reference to Cabot's voyage of 1498, makes the following statement: "It was during this voyage that John Cabot met his demise. His ship sank and was never found. It is presumed that his last place was Grates Cove at the northern tip of the Avalon Peninsula." [1] Peter Firstbrook in his book on Cabot published in 1997 says, "The Grates Cove legend is a fascinating story that is still held dear by people local to the area, but it is given little credence by historians." [2] So what is the background to the story that John Cabot was shipwrecked near Grates Cove on his second expedition in 1498? Why is it that professional historians fail to take it seriously whereas the local people and indeed a few prominent Newfoundlanders have always taken it seriously? Leo E. F. English (1887-1971), a Newfoundland historian, who was born at Job's Cove, Conception Bay not far from Grates Cove has over the years been blamed for encouraging Newfoundlanders to believe the story that Cabot ended his days in Baccalieu Tickle near Grates Cove. English, a former teacher who was later the curator of the Newfoundland Museum in St. Johns, had over the years made many references to the "Cabot Rock." In my opinion English was a kind of propagandist for the Newfoundland

41

tourist industry. He wanted the "Cabot Rock" to be historically
authentic, still he knew that it was only a legend and a very weak argument for Cabot's being near Grates Cove in 1498.

Harold Horwood (b. 1923) in a popular book published in 1969
effectively put words in English's mouth. Here is what Horwood
wrote: "Grates Cove has a well-known rock with some old carvings
on it on a cliff face well above high water. Some of those who formerly examined it, including a curator of the Newfoundland Museum,
professed to be able to read the names, 'IO CABOTO,' 'SANCIUS,'
and 'SAINMALIA' quite plainly." [3] A few weeks after Horwood's
book was published, English "professed" to me that he said nothing
of the sort. In the same paragraph of his book Horwood wrote these
words: "For my part, I am unable to read anything on the rock. It
certainly had some kind of inscription at one time, but I doubt that
anything can be deduced from it today." [4] Near the end of his life,
and even before, English became less and less enthusiastic about the
"Cabot Rock;" yet thirty-five years before his death at the age of 84,
English was writing letters to newspapers and making speeches and
one of his favourite topics was the rock at Grates Cove. In March
1936 he gave a speech probably to the Newfoundland Historical
Society. The St. John's *Daily News* of March 13, 1936 reported the
details of that speech:

> . . . Mr. Leo E.F. English then read a most interesting paper on
> the Cabot Rock at Grates Cove. He took as his text the words of
> the Milanese Ambassador reporting to the Duke of Milan on
> December 18th, 1497, in which the statement occurs that Cabot not
> only took possession and hoisted the Royal Standard, but that he
> also "made marks and returned." These marks, Mr English con-
> tended, are to be found on the rock under the fishing stage of Mrs.
> Meadus on the north side of Grates Cove. He quoted the remarks
> of W. E. Cormack in 1822. Today the inscription cannot be seen
> with the naked eye but the camera reveals parts of it. *Io Caboto* ap-
> pears on one line and the letter *ric* part no doubt of Henricus (i.e.
> Henry VII.) on the next. Objection has been made that an inscrip-
> tion on a rock should be in uncial not script lettering. Mr. English
> pointed out that many inscriptions of the period were in script let-
> ter. Mr. English traced Cabot's course and showed that the mouth
> of Trinity Bay was the logical place near which to look for Cabot's
> landfall and that the large island referred to in the documents of
> that time is, undoubtedly Baccalieu Island. [5]

For many years English had a special photo on display in the
Newfoundland Museum. It was a photo of the rock at Grates Cove,

but not much could be made out of the carvings on that rock; never-theless, certain people claimed that they could make out the words "IO CABOTO." I, myself could never see anything remotely like "CABOTO" on that photo. On October 20, 1955 English published a letter in the St. John's *Daily News* in which he made this statement: "Unless absolute proof is forthcoming, we must maintain that the in-scription at Grates Cove is not a record of Cabot's second voyage. We contend that it is an actual relic of the first voyage of 1497." [6] So English, after all was said and done, was not claiming that Cabot was shipwrecked in Baccalieu Tickle. Rather what he was saying was that Cabot and his men, while ashore in Grates Cove on June 24, 1497, got out chisel and hammer and meticulously engraved some words on the rock. To my mind, the statement above almost gets Leo E. F. English off the hook regarding the legend of the "Cabot Rock;" yet, this Newfoundland historian is not blameless as far as making a lot out of nothing about a few markings on a rock.

So who do we blame for getting the local people worked up about a few markings on a rock on the rough rocky seashore at Grates Cove. The list past and present includes the following: William E. Cormack, Leo E.F. English, Arthur Davies, Harold Horwood, and more recently Arthur Sullivan, Fred Cram and Barbara Shaw who, in 1994 during Grates Coves "Come Home Year," wrote a *Town Crier* speech about Cabot and Grates Cove. [7] I do not wish to appear to be over critical about the subject, but history and legend are not the same thing and a legend should not be purported to be based on authentic historical fact. Hence despite what these people named above believe, there is not one iota of historical evidence to connect Cabot with Grates Cove in either 1497 or 1498.

William E. Cormack (1796-1868), a Newfoundland writer and ad-venturer, wrote a report about his travels across Newfoundland in 1822. In that book which was revised, reprinted and widely dis-tributed in 1873, five years after his death, Cormack claimed that, "The point of Grates is the part of North America first discovered by Europeans." Cormack was confused. He claimed that it was Sebas-tian Cabot who landed at Grates Cove in 1496 (even got the date wrong). Cormack went on to say that, "He (Sebastian or John?) recorded the event by cutting an inscription, still perfectly legible, on a large block of rock that stands on the shore." [8] Cormack, who was an amateur historian, probably never saw the rock at Grates Cove. He was likely influenced by Richard Bonnycastle's map of 1842. On that map opposite Baccalieu Island can be found the following words: "First land discovered by Sebastian Caboto, the land called Prima Vista on June 24, 1497." [9] This might be where Cormack got the idea that it was Sebastian Cabot who landed at Grates Cove in 1497.

Cormack did not claim that Cabot was shipwrecked in Baccalieu Tickle in 1498. Others concocted this story and so a legend was born. The legend is that one of Cabot's ships (the one Cabot was on) in 1498 was shipwrecked in Baccalieu Tickle and some of the crew managed to get ashore at nearby Grates Cove and while awaiting rescue which never came they carved their names and Cabot's name in the rock. The castaways at Grates Cove expected to be rescued by another ship in the expedition which was exploring farther south near Cape Spear. [10]

The rock with the carvings was located close to the old government wharf site in Grates Cove, and a short distance from where a small monument is now located. The old wharf was abandoned in the mid 1950's. There is another rock still standing at the site which looks almost like a twin to the "Cabot Rock" which the local people say was taken away by persons unknown in the 1960's. The fact is the rock was still there in the summer of 1969 when an American historian and myself tried to make some sense out of the markings but to no avail. The rock might have fallen into the water and settled down among dozens of others in the cove. I have never accepted the story that two or three individuals arrived with a truck and simply took the rock away. The legion who built the pyramids of Egypt would be needed to move the rock with the carvings that was once standing in Grates Cove.

Over the years amateur and professional historians have studied the markings on the so-called "Cabot Rock." In 1905 William A. Munn, a Newfoundland historian, could not relate the carvings to Cabot. Neither could I in 1969. In 1969 Samuel E. Morison sent an associate to check out the rock. Raymond Hawtins could find nothing on the rock to resemble "Cabot" or "Caboto." [11] Still, a few Newfoundland writers and the local people in and around Grates Cove have persisted in their claim that the "Cabot Rock" was sure proof that Cabot lost his life near Grates Cove in 1498.

Maybe over the years all this talk about the rock has been done to promote the tourist industry in the area, and this is acceptable so long as people realize that the story of the "Cabot Rock" is a legend and the rock itself like the Plymouth Rock near Boston a perfect fake. Writers like Fred Cram and Barbara Shaw probably pray daily that the time will come when tourists heading towards Grates Cove will be like the Moslems heading towards Mecca.

Those promoting the legend of the "Cabot Rock" had their efforts greatly rewarded in 1955 when Arthur Davies, a British scholar, published an article in *Nature* magazine and gave credence to the story that Cabot ended his days near Grates Cove.[12] Davies, a professor of geography for many years at the University of Exeter in Devon, England, had served in the British army and was with those who landed on the beaches of France during the Normandy Invasion. The story goes that he got to know a couple of Newfoundland soldiers who told him about the "Cabot Rock" at Grates Cove.[13] Ten years later Davies gave a speech on the subject and later wrote his article which is entitled, "The Last Voyage of John Cabot and the Rock at Grates Cove." [14]

Davies knew very little about John Cabot and less about Grates Cove, a place which he never visited. There is a possibility he saw Bonnycastle's map with the inscription opposite Baccalieu Tickle, then suddenly everything about the "Cabot Rock" and Grates Cove was for real. This was very encouraging for those promoting the tourist industry because in time it was hoped Grates Cove would become an international tourist site. But Newfoundland's two best known historians at the time paid little or no attention to Arthur Davies and his Cabot theories. Yet, in a short time, American scholars like Samuel E. Morison, Raymond Hawtins and Allan N. Squires became interested in Grates Cove's piece of "authentic history." Gordon Rothney and Allan Fraser of Memorial University of Newfoundland let it be known that there was no evidence to support Davies' theory. In a letter to Squires in May 1963 Fraser wrote as follows: "This stone is a section of the rock face of a foreshore of the cove and on its vertical face towards the sea, very close to sea level, a few marks or scrapes appear. It is impossible to say definitely whether the marks were man-made or caused naturally . . . There is no historic record or reliable tradition that connects the marks with Cabot." [15]

Although Gordon Rothney and Allan Fraser wrote private letters to scholars on both sides of the Atlantic regarding Davies' speech and subsequent article in *Nature* magazine neither said much publicly and so the "Cabot Rock" legend did not die so to speak. In fact, in time, like a freshly-watered flower it burst forth in bloom and in recent times the "Cabot Rock" legend has received considerable support from local writers and journalists. During and leading up to the Cabot celebrations of 1997, Grates Cove and its legend was not neglected by anybody, and individuals like Fred Cram and Mike Flynn always managed to get items about Grates Cove in *The Compass* or *The Evening Telegram*.[16]

What really, if anything, of significance was on that rock at Grates Cove? In the summer of 1905, the Newfoundland historian, William A. Munn, spent two days at Grates Cove as a guest of Benjamin Benson. At that time Munn copied the full inscription off the rock in order that it would be preserved for all time. The *Newfoundland Quarterly* reported Munn's observations as follows: "His measurements showed the stone to be about seven feet high with the top about four feet and it was carved on three sides. On one side were the letters RW, RH, 1713; IS 1669: IH, AX; on another, VN 1617, N. with a stroke through it, and 1679' HM 1670, BF and B; on the third side EB, IBHM, FC and ILR."[17] I challenge any historian or anyone else for that matter to connect these letters and dates with John Cabot. It just can't be done; so Arthur Davies' theory might have to be laid to rest. The writings carved in the rock that was once standing on the seashore at Grates Cove may have some historical significance but it was not and is not related to John Cabot. Very likely it is nothing more than the initials and the dates of early local explorers. I am aware of similar inscriptions at several places on the Labrador Coast.[18]

On February 8, 1995 a journalist, Mike Flynn, published a short article in the St. John's *The Evening Telegram* datelined Grates Cove. Effectively the article was an interview with Fred Cram, the modern-day chief proponent of the Grates Cove-Cabot theory. Flynn begins his article with these words: "Some people may dispute John Cabot's landing site in the New World, but there should be no question about his final resting place, says a Trinity Bay man." In the interview with Flynn, Cram claims that, "There is sufficient historical documentation to prove the claim" that Cabot was lost near Grates Cove in 1498. Cram then goes on to refer to the, "tremendous amount of research" done by British scholar, Arthur Davies, leading up to the article published in November 1955.[19] If Cram had taken the time to carefully analyse Davies' article, he would have realized that very little research went into the article, three-parts of which is common knowledge about Cabot and the rest guesswork. Davies in his article really only scraped the surface, and there is nothing in it to authenticate the theory that Cabot was lost in Baccalieu Tickle; nevertheless Davies had succeeded in convincing Fred Cram that there is a connection between Cabot and Grates Cove. In 1994 Cram visited Bristol and was surprised to find there two plaques: "One explains why Bonavista is accepted by the British as Cabot's landing site, while the second explains why they believe Cabot spent his final days in the Grates Cove area." [20] Then Cram in his own words admitted that up to that time he was not aware of Davies' comprehensive article.[21] Davies might have believed what he wrote in 1955 and Fred Cram might believe it today, but no professional historian past or present has ever taken Arthur Davies and his article about Cabot seriously.

In the end, Cram's efforts paid dividends and his suggestions were adhered to by the powers-that-be. On June 27, 1997 the *Matthew* visited Grates Cove as one of its ports of call and the crew dropped a wreath in Baccalieu Tickle in memory of John Cabot. What a farce, if ever there was one!

In 1994 the people of Grates Cove and area held a "Come Home Year" and part of the celebrations included the erection of a small monument near where the "Cabot Rock" used to be. The inscription on the monument reads as follows:

THE LEGEND OF THE CABOT ROCK

It is the belief of the residents of Grates Cove that JOHN CABOT landed here in 1497. This belief was fueled by the presence of a large rock high above the water on a cliff in this area. This rock has always been known as the "CABOT ROCK." Some of those who formerly examined the rock including a curator of the "Newfoundland Museum" professed to be able to read the names "IO CABOTO" "SANCIUS" & "SAINMALIA" quite plainly. It is certain that in the mid 1960's two men from the media removed the face of the rock which contained these inscriptions.

Erected August 14, 1994
COME HOME YEAR[22]

Also as part of the ceremony associated with the unveiling of the monument a *Town Crier* in a well-written speech announced to the world the connection between Grates Cove and John Cabot.[23] The speech written by Barbara E. Shaw is well put together but reflects the same ideas and in some places used the same words and phrases as found in Arthur Davies' article. In fact, Shaw, whether knowingly or unknowingly, makes the same spelling mistakes in her *Town Crier* speech as Davies makes in his article; for example, using HAWLEY for HOWLEY with reference to Archbishop Howley (1843-1914), a Cabotian scholar, and the Roman Catholic archbishop of St. John's a century ago.[24] History records that there was nobody living at Grates Cove before 1790; hence, the notion that, "it is traditionally believed" that Cabot was at or near Grates Cove in either 1497 or 1498 is pure speculation.[25] Speculation is no substitute for historical evidence.

Notes to Chapter 5

The "Cabot Rock" at Grates Cove

1. Francesco Alusio, *Giovanni Cabot (John Cabot): A Passion for Discovery* (Mississauga, Ontario, 1997), p. 46.

2. Peter Firstbrook, *The Voyage of the Matthew: John Cabot and the Discovery of North America* (Toronto, 1997), p. 137.

3. Harold Horwood, *Newfoundland* (Toronto, 1969), p. 161.

4. *Loc. cit.*

5. *The Daily News*, St. John's, Newfoundland, March 13, 1936.

6. *The Daily News*, St. John's, Newfoundland, October 20, 1955.

7. Barbara E. Shaw, *The True History of Giovani Cabot: Grates Cove and the Cabot Rock,* (privately printed), Grates Cove, 1995.

8. William E. Cormack, *Narrative of a Journey Across the Island of Newfoundland* (London, 1822). Reprinted in the *Morning Chronicle*, St. John's (September, 1873).

9. See Richard Bonnycastle's map of Newfoundland, 1842. Portion of map reproduced in John Parsons, *Away Beyond the Virgin Rocks* (St. John's, 1997), p. 224.

10. Arthur Davies, "The Last Voyage of John Cabot and the Rock at Grates Cove," *Nature*, Volume 176, November 26, 1955, pp. 996-999. See also Shaw, *op. cit.*

11. Samuel E. Morison, *The European Discovery of America: The Northern Voyages, 500-1600 A.D.* (New York, 1971), p. 203.

12. Davies, *op. cit.*

13. Personal knowledge from Richard Parsons, Graham Bursey, Thomas G. Mercer, and Alfred Blundon.

14. Davies, *op. cit.*

15. Allan Fraser, *Personal Papers*, Provincial Archives of Newfoundland. See also *The Compass*, July 15, 1997.

16. See *The Evening Telegram*, February 8, 1995, *The Express*, December 4, 1996 and *The Compass*, July 1 and 15, 1997.

17. See *The Newfoundland Quarterly*, December, 1955, Volume 54, No. 4, p. 48.

18. John Parsons, *Away Beyond the Virgin Rocks: A Tribute to John Cabot* (St. John's, 1997), p. 23.

19. See *The Evening Telegram*, February 8, 1995.

20. *Loc. cit.*

21. *Loc. cit.*

22. Cabot Monument, Grates Cove. See *The Compass*, July 15, 1997.

23. Barbara E. Shaw, *op. cit.*

24. See Shaw, *op. cit.* and Davies, op. cit.

25. See item on Grates Cove, *Encyclopedia of Newfoundland and Labrador*, St. John's, Volume 2, p. 700.

Fred Cram, a Newfoundland educator and writer who, like Arthur Davies, believes that Cabot spent his last days in the Grates Cove area.

— The Author's Collection

Chapter 6

Arthur M. Sullivan's Letter and Arthur Davies' Article

Arthur Davies' article in *Nature* magazine has over the years not received much attention from historians or anybody else. Initially, the article caused some excitement in some academic circles mainly because of Davies' reputation as a British scholar, but letters by Allan Fraser and Gordon Rothney to scholars about Davies' article dampened some people's feelings, even those of local political propagandists who were anxious to have anything of a historical nature that would enhance the local tourist industry. Hence, for over forty years now, only Samuel E. Morison and Allan Fraser among historians have come forth and condemned Davies theory outright.[1] Recently Peter E. Pope, Peter Firstbrook and myself have been mildly critized for our negative attitude towards Arthur Davies' theory and the whole matter of the connection between Cabot and Grates Cove. Pope considers the legend of the "Cabot Rock" as a kind of weak argument for claiming that Cabot was ever in Grates Cove, and Firstbrook says that the fascinating story of the Grates Cove legend, "is given little credence by historians." Alan F. Williams in his book on Cabot calls Davies' theory, "highly speculative," and in my own writings, I have always claimed that the "Cabot Rock" story is a good Newfoundland tall tale.[2] The fact is that over the past forty years Arthur Davies and his theory about Cabot and Grates Cove have virtually been ignored by almost everybody of academic or literary importance particularly professional historians. Williamson in his book in 1962 doesn't mention Davies and neither does Wilson in his popular book, *The Columbus Myth* in 1991. Raleigh A. Skelton in his essay on cartography in Williamson's book makes a passing reference to Davies, but doesn't mention Davies at all in his scholarly article on

50

John Cabot in the *Dictionary of Canadian Biography* (1966). Davies' theory is not mentioned in any encyclopedias or historical dictionaries. But people like a good story and sometimes try to turn legends into truth, so that over the years local people have argued and still argue about and write about Grates Cove and its "Cabot Rock" as if it were for real.

The chief proponent of the "Cabot Rock" story in Newfoundland, Fred Cram, educator and writer, who lives in Old Perlican near Grates Cove, by his own admission to a local journalist only discovered the Davies' article on a trip to England in 1994. Cram apparently was greatly impressed because Davies provided the "scholarly" evidence that Cram and others needed to promote a legend as a fact. [3] In late 1994 a retired university professor, writer and tour guide director, Arthur M. Sullivan, came forward as the latest proponent of the "Cabot Rock" story. Sullivan too had come upon the Davies' article and he too, like Cram, was very impressed by what Arthur Davies wrote in 1955.

Reposing in the Cabot collection of newspaper clippings, reports, documents, book extracts, and speeches that make up the *Cabot Biographical File* at the Centre for Newfoundland studies at Memorial University of Newfoundland in St. John's is a two page letter officially signed by Arthur M. Sullivan and addressed to Miller Ayre, chairman at the time, of the John Cabot (1997) 500th Anniversary Corporation. The letter is dated December 19, 1994. Sullivan apparently had sent the letter to Anne Hart, Head of the Centre for Newfoundland Studies, with a request that it be placed in the *Cabot Biographical File*. The fact that the letter was filed in a library indicates that there was nothing secretive or private about its contents. Sullivan who may have had some contact with Cram felt the Davies' article was all that was needed as support for the people of Grates Cove in their efforts to have the *Matthew* call at their community in 1997. The *Matthew* did indeed stop at Grates Cove in 1997. One hundred years from now Sullivans letter will be considered a significant document in relation to the Cabot 500th celebrations of 1997. In fact, it would be argued that it is already a significant document.

Arthur M. Sullivan, a former Newfoundland Rhodes scholar (1957), follows a number of prominent individuals who over the years have advocated the legitimacy of Arthur Davies' theory that Cabot ended his days in Baccalieu Tickle near Grates Cove. Yet, with the exception of Davies, all those who have attempted to turn a legend into historical fact are local and among others the list includes William E. Cormack, Leo E. F. English, Harold Horwood, Fred Cram, Barbara E. Shaw and Eleanor Butt. Arthur M. Sullivan is the most recent to advocate that the "Cabot Rock" story on the basis of Davies' article be no longer regarded as a legend, but rather regarded as a true historical interpretation as to what happened to John Cabot on his mysterious voyage of 1498.

To my mind, Sullivan's letter is so important that I dare not comment on it unless its contents are revealed so that readers can form their own opinion as to whether or not it has any legitimacy, and can be regarded as a valid argument in support of Davies and the Grates Cove "Cabot Rock" story. Here then is Arthur M. Sullivan's letter in full:

Dear Miller,

Some days ago, I spoke with you on a CBC phone-in show and passed on the suggestion that in 1997 the Matthew should drop a wreath in Grates Cove since that is one of the three most likely places for John Cabot to have died. At that time the only supporting evidence that I had was a plaque at the site of the Matthew reconstruction in Bristol.

Since then I have, with the assistance of the Center for Newfoundland Studies, found several documents which make direct reference to the Cabot stone in Grates Cove and the second voyage of Cabot in 1498. I enclose the papers for your perusal. The most scholarly and impressive is by Prof. Davis of Exeter University. He makes a forceful and persuasive argument that Cabot did, in fact, die at, or near, Grates Cove in December of 1498. His entire article contains much interesting and relevant material, but his argument concerning Cabot's death may be summarized as follows:

1. Following his 1497 voyage, Cabot was granted a pension of 20 pounds by King Henry VII - to be paid half yearly. One payment was made in the Fall of 1498 and a second on March 21, 1499. No further payments were made. There must, therefore, have been clear and compelling evidence that Cabot was not alive/available before the third payment was due in the Fall of 1499. But there is no trace of John Cabot's death or burial in Bristol or indeed in the records of any part of England.

2. In 1498 John Cabot set sail for a second voyage to the new founde land. He was in command of five ships. He took sufficient provisions for one year so he obviously did not intend to return until the summer of 1499. From the sketchy evidence available, it appears that one ship turned back before clearing the coast of Ireland, two others did not reach the Newfoundland waters at all, and of the two to reach the Newfoundland waters (Cabot's ship

and a consort vessel), only one, the consort vessel, returned to England early in the summer of 1499. This vessel reported the death of John Cabot.

3.   The tragedy appears to have occurred as follows. After fishing for cod around Baccalieu island, the two vessels were separated in a storm. John Cabot's vessel was probably driven ashore at, or near, Grates Cove. The crew did not all perish immediately, and some at least made it ashore where they survived for some time. The other vessel, not having seen the shipwreck, assumed that the two would meet again and continued on to Cape Spear. By this time they realized that something must have happened and retraced their route back to Baccalieu but apart from wreckage, which clearly indicated a disaster, they found no surviving crew members. The vessel then turned back to England on December 8.

4.   In 1501, Gaspar de Corte Real visited the shores of eastern Newfoundland. He captured some 50 natives (Beothucks), one of whom was in possession of a broken sword of Italian origin, and a boy who was wearing earrings of Venetian origin. Since the only other European voyage to this area before 1501 was that of John Cabot, these items could only have come from Cabot's 1498 voyage.

5.   In Grates Cove, until recently, there existed a curiosity, the Cabot stone. This rock was first described by Cormack in 1822 and was subsequently visited by many historians. Leo F. English stated that the inscription read To Caboto, Sancius (the name of Cabot's 14 year old son) and Sainmala. (Unfortunately the part of the rock on which the words were inscribed disappeared under mysterious circumstances during the 1960's and can no longer be seen).

6.   Davis concludes that "one can only assume that Cabot and his party, after inscribing the stone were overcome and killed by Beothuck Indians who acquired Cabot's sword and Sancius' earrings, the items which were later recovered by Corte Real."

As you can see from the attached material and the sequence noted above, the possibility that John Cabot did die in Newfoundland is not entirely fanciful. Perhaps the possibility and its implications for the Cabot year celebrations should be given further consideration.

Sincerely,
Arthur M. Sullivan [4]

As can be seen Sullivan's letter is self-explanatory; nevertheless, a few comments are in order. One wonders what Sullivan has in mind when he writes, "since that [meaning Grates Cove] is one of the three most likely places for John Cabot to have died." Where are the other two likely places? Sullivan, too, like Cram, admits that it is only recently that he has been aware of Arthur Davies article in *Nature* magazine, and before that "the only supporting evidence that I had was a plaque . . . in Bristol." It must be noted that a plaque cannot be taken as supporting evidence for anything, neither can a statue. The fact that there is a statue of Cabot at Bonavista means nothing in terms of history. The fact is, most professional historians nowadays do not regard Cape Bonavista as Cabot's landfall in 1497.

Sullivan then goes on to say that, with the assistance of the Centre of Newfoundland Studies, he has "found several documents which make direct reference to the Cabot stone in Grates Cove and the second voyage of Cabot in 1498." The most impressive of these documents he says is the "most scholarly and impressive" article by Arthur Davies who in Sullivan's opinion "makes a forceful and persuasive argument that Cabot did, in fact, die at, or near, Grates Cove in December 1498." I have no idea what documents Sullivan is referring to unless it is the writings of William Cormack, Leo. E. F. English and a few local journalists over the years; and, if that is the case, it is certainly debatable as to whether or not this material can be called documents. There are no documents — certainly not contemporary documents — anywhere that make any reference to John Cabot ever being near Grates Cove in 1497, 1498 or any other time. In fact, Davies' whole argument or reconstruction is nothing more than an attempt to connect Cabot's voyage of 1498 to an early sixteenth-century map of the North Atlantic area drawn by the Portuguese cartographer, Pedro Reinel. Most historians feel that Davies in 1955 failed to prove anything; and that is precisely why historians as far as the Grates Cove legend is concerned, have all but ignored Arthur Davies over the past forty years or so.

Sullivan who spells the British geographer's name incorrectly, then goes on to summarize Davies' article under six headings. Sullivan's attempt to be persuasive cannot be denied, but there are statements in this letter that are not only inconsistent but factually incorrect. Sullivan's knowledge of what was written on the "Cabot Rock" is obviously second-hand information. We shall continue to analyze the letter. Under heading number 2 Sullivan makes the same mistake that Davies makes in his article — he invents history: "From the sketchy evidence available, it appears that one ship turned back before clearing the coast of Ireland, two others did not reach the Newfoundland waters at all, and of the two to reach the Newfoundland waters (Cabot's ship and a "consort" vessel), only one, the consort vessel, returned to England in the summer of 1499. This vessel reported the death of John Cabot." Apart from the mention that one ship returned to Ireland, everything else in this statement is nothing more than an attempt at historical reconstruction which is based on absolutely no evidence. Effectively, Sullivan's argument is flawed because he, too, is guilty of the sin of historical assumption and is merely echoing what Davies says in his article. No authority on Cabot would go along with the above statement. In other words, there is no historical evidence to support what Sullivan says under heading number 2 other than the reference that one ship returned to an Irish port. If only Sullivan had read James A. Williamson's book (1962), the recognized authority in Cabotian studies these kinds of historical suppositions could have been avoided. The statements that two ships made it to Newfoundland and, "two others did not reach the Newfoundland waters at all" and the two that did arrive in New-foundland were Cabot's ship and a "consort" vessel, is a fairy tale. There is not an iota of historical evidence anywhere to support it other than what is contained in Davies' article written in 1955 and Davies is not a histoically recognized source.

Under heading number 4 we find this great historical revelation: "Since the only other European voyage to this area before 1501 was that of John Cabot, these items could only have come from Cabot's 1498 voyage." I agree with this statement in that it fits the evidence of the contemporary documents, but it can be argued that the items referred to could have come from the voyage of 1497, or even from the Bristol fishermen who were known to have fished around what is now the Avalon Peninsula (their Isle of Brasil) at least two decades before Cabot's first voyage.^{5} Also there is no evidence that the items in question — the piece of broken sword and the earrings — were found in Newfoundland. They could have been found in Nova Scotia and the Indians referred to in the original document (dated October 19, 1501) could have been Micmacs. In heading number 5 Sullivan implicates the late Leo E. F. English once again. As we have seen earlier, English in 1955 in a letter to the St. John's *Daily News* stated publicly that it was his belief that the engravings on the rock at Grates Cove, "is not a record of shipwreck on Cabot's second voyage." In fact, in the same letter English wrote these words: "We contend that it is an actual record or relic of the first voyage of 1497."[6] By 1969 English had changed his mind completely and claimed that in his judgement the rock at Grates Cove had no refer-ence to John Cabot or any other historical event going as far back as 1497 or 1498.[7]

In heading number 6 Sullivan uses a direct quotation from Davies' article: "One can only assume that Cabot and his party, after inscribing the stone were overcome and killed by Beothuk Indians who acquired Cabot's sword and Sancius' earrings, the items which were later recovered by Corte Real." No wonder the Newfoundlanders killed off the Beothuks because, according to Davies, they were the perpetrators of the most vicious and insidious crime in the history of Newfoundland. They were the ones who killed John Cabot the great Italian navigator who discovered North America. So the Beothuks were the ones who took Cabot's sword and Sancius' earrings, then headed up the Bay de Verde peninsula, through the isthmus of Avalon, and back to the shores of the Exploits River, their natural habitat. The key word to notice in Davies' quotation in Sullivan's heading number 6 is the word "assume." Sullivan ends his letter to Miller Ayre by referring the latter to the material attached and also stated that, "the possibility that John Cabot did die in Newfoundland is not entirely fanciful." The evidence for this, of course, being the article by Arthur Davies which is no evidence at all.

It is doubtful as to whether Davies even checked out the contemporary documents relating to Cabot's voyage of 1498. Yet Sullivan's letter was a calculated attempt to support the efforts of Fred Cram, Barbara E. Shaw, Eleanor Butt and others to get the *Matthew* to make Grates Cove one of its ports of call in June 1997. In that regard Sullivan's letter had the desired effect; still from a historical perspective any attempt to support Arthur Davies, and thus give credence to his theories as set forth in his article in *Nature* magazine in 1955 is nothing more than as act of farcical proportions. No reputable historians have ever swallowed Arthur Davies' bait.

The foregoing reference to Arthur Sullivan's letter and the references to Arthur Davies' article might encourage a few readers to head for the nearest public or university library and check out Davies' article published in *Nature* magazine, Volume 176, Number 4491, dated November 26, 1955. Since it is unlikely that many readers will bother to check it out, a few comments from me on Davies' article are undoubtedly in order here. So let us take a close look at the article entitled, "The Last Voyage of John Cabot and the Rock at Grates Cove" and see does it stand up to critical historical analysis. Let us find out what Davies said in 1955 which so impressed Fred Cram and Arthur M. Sullivan over forty years later.

Davies who on September 6, 1953 had given a speech on Cabot at Bristol before Section E (Geography) of the British Association apparently decided sometime in the next couple of years that this speech was worthy of publication, and so on November 26, 1955 his mini-dissertation on Cabot and his connection with Grates Cove turned up in the pages of *Nature* magazine. What does Davies really say in the famous or infamous article? The truth is that he does not say much; in fact, he says so little that he adds next to nothing to the Cabot debate, and his influence on other British scholars particularly Cabotian scholars was next to nil. As I have indicated, the British cartographer, Raleigh A. Skelton, makes only a passing reference to Davies in his essay on cartography in Williamson's book, and Williamson, himself, does not even mention Davies.

In the first half of his article Davies gives background information that every Cabotian amateur scholar is familiar with, and then seemingly obsessed with the piece of broken gilt sword and silver earrings, he starts assuming things; but assumptions cannot be regarded as valid historical knowledge and Davies, a professionally trained individual, should have known that. Historical interpretations can only be made from an evaluation and assessment of contemporary documents whether literary or cartographical. Davies bases his whole thesis on one map drawn by a Portuguese cartographer. The map in question shows eighteen place names, some of which Davies claims had Cabot associations and represented a sequence of saints' days.[8] But Davies is at his best when he moves into the realms of possibility, probability and assumption. Witness the following statement about the broken sword and the silver earrings with reference to Corte Real's voyage of 1501:

> At some place, probably in east Newfoundland, fifty Indian prisoners were taken, one of them being in possession of a broken gilt sword of Italian origin (which points to John Cabot), while an Indian boy was wearing silver earrings of Venetian origin (which points to Sancius, the young son of John Cabot, about fourteen years of age). The only voyage in these waters previous to that of Corte Real was the 1498 voyage of John Cabot. These relics are mute witnesses that John Cabot and Sancius either reached shore alive or that their bodies were washed ashore. Indian custom would ensure that an Indian boy wore the earrings only if they were found on the boy Sancius; otherwise they would have been worn by a man.[9]

The Indian prisoners could have been taken in Nova Scotia or New England. Also there is no document which says that Cabot's sons were with him on the voyage of 1498 or 1497. The reference to the relics being mute witnesses is purely assumption on Davies' part, since most reputable scholars believe the Indians acquired these items in a trading transaction. Finally, note in the quotation above that Davies is even versed in "Indian custom" of the time. Davies goes on to say that, "This Reinel map of 1502 corresponds exactly with the discoveries claimed by Sebastian Cabot, thus proving that the voyage which made those discoveries had been completed before 1502. It rules out any possibility that Sebastian could have led it, for he was then less than twenty years of age." [10] There is no indication in any contemporary document that anybody from the Cabot expedition of 1498 returned to England. How could Sebastian know about the discoveries of John Cabot's voyage of 1498 when he was not on that voyage?

Throughout his article Davies keeps making statements that seem to be based on fact; that is, as if they can be historically documented, but this is not really the case. Most of Davies' statements are assumptions and suppositions that just do not mesh with even the most liberal interpretation of the documents. Other well-known historians and cartographers have over the years studied the contemporary Cabot documents but nobody, to my mind, has assumed anything unless there is some evidence in the documents to justify an assumption. Most of Davies' article is based on assumptions and presented in such a way as if no contemporary documents exist. Williamson in his classic study of Cabot (1962) includes forty documents related to Cabot's voyages of 1497 and 1498. [11] Davies also implies that the general outline of the La Cosa map was brought back to England by the surviving ship of the Cabot expedition of 1498: "The general outline of the 1498-99 Cabot voyage, brought back by the surviving ship to Henry VII, may well have been copied roughly by the Spanish ambassador and sent to Spain, as had been done in 1497." [12] How can this be so when, according to the best documented evidence, no ship from the ill-fated expedition of 1498 returned. Davies even says that, "priests never went on dangerous voyages of exploration, only on later voyages of colonization and settlement," yet no less than two documents state clearly that there were two priests on Cabot's voyage of 1498. [13]

Arthur Davies' article is nothing more than an attempt to create a sequence of events that would seem to give logical explanations regarding Cabot's voyage of 1498. In other words, Davies, like Newfoundland amateur historian, Jack Dodd (1902-1978), tried to write history the way he thought it happened without any reference to contemporary documents.[14] Such reconstructions are, of course, suspect and highly speculative and in the end are good for amateur debates but do nothing to enhance historical knowledge. At this point Arthur Davies is allowed to speak for himself:

> Baya de Santa Cyria indicates August 8. Y. dos bacalhaos is island of cod fish. Sebastian Cabot told Peter Martyr that he had given this name to east Newfoundland because of the plentiful supply of cod fish found there; but Martyr noted that there were many Spaniards who denied that Sebastian Cabot had sailed so far west and discovered this land. Baya do concepcam indicates the feast of the Conception on December 8, which points to a very long delay in this region. The next name, C. da espera, gives the explanation of this long delay, for it means the cape of waiting. To English ears espera sounds like spear, and Cape Spear occurs on modern maps just south of Saint John. It is possible that John Cabot took only two ships on this westward exploration, the other three ships which sailed out in company from Bristol being bound for Iceland. Thus there would be one consort ship, and it is probable that this ship set up a look-out post at C. da espera and then spent months searching the coasts and rivers to the south and then to the north for signs of survivors. This was the duty of a consort ship, and faithfully they pressed their fruitless search month after month. R. de sam francisco to the south is October 4, while baya do concepcam to the north is December 8. The tragedy would seem to have occurred as follows. After fishing for cod fish near the island of Bacalhaos (the modern Baccalieu island), they turned south to cross Conception Bay. Possibly during hours of darkness the ship of John Cabot "went down in the depth of the ocean being seized by the ocean itself." The description almost implies that it was not wrecked on a coast nor sunk by tempest. "Being seized by the ocean itself" suggests that the consort ship was at a loss to account for the disappearance of the Cabot ship. It is possible that the ship carrying John Cabot struck an iceberg during darkness not far from Baccalieu Island, at the north end of Conception Bay. Near by is Grates Point, and just west of it is Grates Cove. It may be that Cabot, his young son Sancius and some of the crew succeeded in reaching Grates Cove in the ship's boat. Cabot and Sancius certainly got ashore, alive or dead, for the broken Italian gilt sword points to Cabot himself and the Venetian silver earrings to the boy Sancius.[15]

Notice that Davies used "Saint John" when he meant "St. John's," "It is probable," "The description almost implies," "It may be," "When it became apparent," "They may have discovered," "most probably," and a few others are used throughout the extract as is the case with the whole article. Davies in his article also refers to a document that confuses "John Cabot" with "Antonio Gaboto" but he does not attempt to enlighten his readers with an explanation.

Davies in the course of his reconstruction of events pertaining to John Cabot and Grates Cove continues his ridiculous reconstruction as follows:

> An important piece of evidence in this reconstruction of Cabot's last days is the large rock at Grates Cove, near Grates Point, north of the headland which terminates Conception Bay and practically opposite Baccalieu Island. This rock, which was first described in 1822 by William Epps Cormack as "a large block of rock which stands on the shore," bears an inscription, still partly discernible, in which the words Jo Caboto can be plainly traced. Parts of other names are legible, and among these are Sancius and Sainmalia. Leo F. English, curator of the museum at St. John's, Newfoundland, has recently commented on this rock; he quotes Archbishop Hawley for the statement that Grates Cove comes from the Italian gratia, thanks be to God. No English names can be deciphered.
>
> On the supposition that John Cabot and Sancius with some of the crew got ashore at Grates Cove, it would seem that, knowing the consort ship would search the coasts for them, they inscribed these names on the rock at Grates Cove. No one by the name of Sainmalia is known to be connected with the expedition. It may be that it is the illegible remnant of "Santa Maria save us" or some such expression. It is not credible that such an inscription could be anything but genuine, for no forger before 1822, however learned, knew of the Reinel 1502 map, which was first discovered in 1844, and not until today has anyone reconstructed the 1498 voyage of John Cabot and suggested that his ship went down in the vicinity of Grates Cove. One can only assume that Cabot and his party, after inscribing the stone, were overcome and killed by the Beothuk Indians. The searchers from the consort ship never went north of Baccalieu Island and never saw the inscription on the rock. Had they done so, it is possible they might have saved John Cabot and his men, stranded on this wild coast to end their days.
>
> The rock at Grates Cove, the broken gilt sword of Italian origin and the silver earrings of Venetian origin are relics of the disaster which overtook John Cabot. With the aid of Portuguese maps of America they help to reconstruct in detail the fortunes and achievements of the 1498-99 voyage, one of the greatest triumphs over hardship and danger that exists in the annals of American discovery.[16]

A discerning reader will notice that Leo E. F. English once again enters the picture, that Archbishop Howley's name is spelled incorrectly, and that expressions like," It may be," "It is not credible," and "One can only assume," are used in this extract.

One is at a loss to understand why Fred Cram and Arthur M.

Sullivan both reputable scholars would be impressed by this article. It is not a scholarly article and there is nothing impressive about it; in fact, it is nothing more than an attempt to relate John Cabot and Grates Cove by presenting a historical reconstruction based on precious little evidence. This would be a perfectly logical explanation of what happened to Cabot if there were historical evidence to support it. But there is not; and history becomes nothing more than a tall tale or a folktale unless there is supporting evidence.

Ethel S. Patterson, a British geographer, in a tribute to Davies when he retired from the University of Exeter in 1971, wrote the following relative to Davies' interest in the Cabot voyages to North America:

> His dual interest in the scientific and human factors in his studies made him give full weight not only to the scientific interest in the voyages, but also to the lives and characters of the men who made them. One feels, as one reads his work, that he was not only writing about these journeys, but also adventuring along with the men who made them.[17]

Arthur Davies is recognized as one of the greatest British geographers of the twentieth century[18], but had this man devoted his life to the writing of fiction he may have become one of the world's great writers. Professional historians must at all times be prepared to differentiate between history and fiction. Davies did not do this, and this is precisely why over the years historians have paid little attention to his attempt to relate John Cabot with Grates Cove and its rock.

David Alan-Williams[19], who in 1997 commanded the *Matthew* (the replica of Cabot's ship on his first voyage) from England to Newfoundland, in the course of his remarks at Grates Cove on June 27, 1997 made this statement: "I'm glad to say I didn't repeat the history of 499 years ago." [20] He was referring specifically to John Cabot's being shipwrecked in Baccalieu Tickle in 1498. David Alan-Williams, world-famous navigator and scholar, has something in common with Arthur Davies. He, too, cannot distinguish between fact and fiction.

Notes to Chapter 6

Arthur M. Sullivan's Letter and Arthur Davies' Article

1. Samuel E. Morison, *The European Discovery of America: The Northern Voyages, 500-1600 A.D.* (New York, 1971), p. 203. Allan Fraser, Personal Papers, Provincial Archives (Newfoundland). See also *The Compass*, July 15, 1997 and John Parsons, *Away Beyond the Virgin Rocks* (St. John's, 1997), p. 21.

2. See Parsons, *op. cit.*, chapter 3.

3. See *The Evening Telegram*, St. John's, February 8, 1995.

4. The Sullivan Letter, *Cabot Biographical File*, Centre for Newfoundland Studies, Memorial University of Newfoundland, St. John's, Newfoundland.

5. Tryggvi J. Oleson and William L. Morton, "The Northern Approaches to Canada," *The Dictionary of Canadian Biography*, Volume 1 (1000-1700), (Toronto, 1966), p. 19-20.

6. Leo E.F. English in *The Daily News*, St. John's, Newfoundland, October 20, 1955.

7. Personal interview with Leo E. F. English, May 18, 1969.

8. Alan F. Williams, *John Cabot and Newfoundland* (St. John's, 1996), pp. 47-48.

9. Arthur Davies, "The Last Voyage of John Cabot and the Rock at Grates Cove," *Nature*, Volume 176, November, 1955, p. 997.

10. *Loc. cit.*

11. James A. Williamson, *The Cabot Voyages and Bristol Discovery under Henry VII* (Cambridge, England, 1962), pp. 175-234.

12. Arthur Davies, *op cit.*, p. 998.

13. Williamson, *op. cit.*, Documents, Numbers 36 and 37.

14. John Parsons, *op. cit.*, See Appendix 20 — Jack Dodd.

15. Arthur Davies, *op. cit.*, p. 998.

16. *Ibid.*, p. 998-999.

17. Ethel S. Patterson, "Professor Arthur Davies: A Memoir" in K. J. Gregory and W.L.D. Ravenhill, (Eds.), *Exeter Essays in Geography in Honour of Arthur Davies*, (Exeter, Devon England, 1971), pp. XIII-XIV.

18. *Loc. cit.*

19. For Biographical Information on David Alan-Williams see "The Matthew Voyage — the old and the new," *Round the Bay — Cruising Newfoundland and Labrador*, 1997.

20. Quotation from Bill Bowman, "Matthew visit to Grates Cove brief, but meaningful stopover," *The Compass*, July 1, 1997.

Epilogue

Historians and writers must make every attempt in their historical writings to be accurate, although at times it is easy to fall into a trap and keep adding to what could be called historical gossip. Anybody today writing about Cabot's second voyage must examine everything written about that subject in the past and then check it against available maps and contemporary literary documents. All secondary materials must be sorted in terms of what is reliable. Information in encyclopedias and some biographical dictionaries is, by and large, based on secondary sources and in some cases is not reliable.

In terms of Cabot's second voyage of 1498, the most reliable secondary sources which are easily accessible are the books by Williamson (1929 and 1962), Wilson (1991 and 1996), Firstbrook (1997), Cuthbertson (1997), and Williams (1997). Raleigh A. Skelton's essay on John Cabot in the *Dictionary of Canadian Biography* (1966) is possibly the best reliable summary anywhere about John Cabot. Historians such as those just mentioned have evaluated their research material, made their interpretations of the documents, and then written their books, the end result being as close to what they think is historical truth at that point in time. Several books in the last hundred years have produced contemporary documents related to the Cabot voyages; the best collections, however, are found in Biggar (1911) and Williamson (1962). Peter Firstbrook, in his recent book on Cabot (1997), also reproduced a number of important Cabot documents. The John Day letter can be found in my book, *Away Beyond the Virgin Rocks* (1997) exactly as it was translated from the Spanish by Louis-Andre Vigneras in 1956.[1]

James A. Williamson's classic study (1962) is the most authoritative work on John Cabot and anybody writing anything about the Italian explorer should not put pen to paper without first examining carefully Williamson's book — and not only Williamson's own writings, but also the documents reproduced in that book as well as Raleigh A. Skelton's excellent essay on the cartography related to the Cabot voyages. This, too, is included in Williamson's book. One hundred years from now this book might be considered out-of-date, but that will only be so if, in the next century, new evidence is discovered which will add to our knowledge of Cabot. A good example of new and valuable evidence is the John Day letter discovered by Louis-Andre Vigneras in a Spanish archives in 1956. It can now be said that Henry P. Biggar's book (1911) is out-of-date, that being so because in its pages one cannot find the John Day letter discovered forty-five years after Biggar's book was published. Hence, in terms of historical subjects like the Cabot voyages, nothing is absolutely final and for certain; new information is always pending.

Arthur Davies' article on John Cabot's connection with Grates Cove can never be considered as a reliable contribution to history simply because most of it is an invention of history — a reconstruction in which the author is more concerned with concocting a story than presenting historical truth. Davies in his article pays little or no attention to all the work done by Cabotian scholars before his time. He builds his whole case around one map and a Newfoundland folktale, and in the end really adds nothing to our knowledge of John Cabot and his two voyages. Some would call it being historically irresponsible and Davies would certainly be guilty of that, but Arthur Davies was scholar enough to know exactly what he was doing. At least he gave the people in and around Grates Cove a boost and he certainly inspired individuals like Fred Cram and Arthur M. Sullivan. From an article about Davies written by a colleague one gets the impression that he was one of the finest of men, and that may be so; nevertheless, in my opinion, he took undue liberties with the truth in his 1955 article.[2] Maybe it was all deliberate, for Davies was well aware that reputable historians in the field of Cabotian studies would put and keep the record straight.

Even reputable historians can be irresponsible if they write books based only on the secondary sources and pay little attention to the primary documents. That is what Henry P. Biggar (1872-1938), the Canadian historian, did in his book, *The Voyages of the Cabots and the Corte-Reals, 1497-1503* published in Paris in 1903. It is true he used a few documents, but most of his work is based on a careful study of sixteenth century secondary source material wherein he made good use of the works of Hakluyt, Eden, Martyr and Ramusio. He also made good use of a work by Antonio Galvano, *The Discoveries of the World,* published in Lisbon in 1731. Biggar cannot be excused for this because all the Cabot documents were available to him except the John Day letter. Because writers concerned with Cabot during the first half of the sixteenth century depended on oral history supplied mainly by Sebastian Cabot, much of what was reported was a combination of lies, half-truths, and confused accounts that went into the record and eventually was regarded as valid history; yet, even a historian of Biggar's calibre, a man Williamson became frustrated with on several occasions, was taken in by all this. In his book, Biggar gives a beautiful account of Cabot's second voyage and his return to Bristol but, according to the latest research, there is hardly a word of truth in what he wrote. Cabot's second voyage is like dying in that nobody ever came back to tell the story; still, according to Biggar, Cabot did come back, but his reception was not too enthusiastic. Here is what Biggar wrote in 1903:

> Of the reception accorded the Cabots on their arrival in England we know little, but it is not difficult to understand that it was not over-enthusiastic. They had set off promising to bring home heavy cargoes of spices and oriental gems. Here they were at last with their ships completely empty of everything. The faces of the shareholders, from the king down, must have fallen considerably. Nor had the Cabots anything of interest to relate. They could merely tell how they had visited a region covered with ice and snow in the north and in the south of bleak, rock-bound coasts like those of Labrador, Newfoundland and New England. Even in a region as pleasant as New Jersey or Maryland no signs had been seen of eastern civilization. Such an account must have dampened considerably the ardour of all who had hitherto been interested in the undertaking. The result was that no fresh expedition was sent out until the year 1501 and that one was not in charge of the Cabots.[3]

Notice that even as late as 1903 one of Canada's best-known historians still believed that Sebastian Cabot was on the voyage of 1498. Also a discerning reader will notice that the above passage written by Biggar is the source for a similar passage found in Arthur P. Newton's book published in London in 1932. Newton was a reputable British historian at King's College, London in the 1920's.[4] Please refer back to the Prologue.

Every historian, besides examining secondary source material carefully, must have the ability to analyze and interpret contemporary documents. The secondary sources must always be evaluated in the light of original documents. Only in this way can reliable information be presented about any historical subject. Some of the material written over the years about John Cabot's voyages have been very unreliable, and in a lot of cases much guesswork has been employed.

In 1914, two Canadian historians, George M. Wrong and Hugh H. Langton edited a series of Canadian textbooks known as the *Chronicles of Canada*. Stephen Leacock, the well-known Canadian humorist, who was also a reputable Canadian scholar, wrote the book which includes the accounts of Cabot's two voyages. Some of the things Leacock wrote about the Cabot voyages as late as 1914 is pure fiction — beautiful writing but terrible history. Not only that, in 1914 Leacock had access to all the Cabotian documents except the John Day letter; in fact, H. P. Biggar had published all the documents in a book in 1911. As well Leacock had access to the writings of reputable Cabotian scholars like Justin Winsor, Charles Deane, Henry Harrisse and Henry P. Biggar. Yet in reality, Leacock cannot be blamed for inventing history, because the sources he used reflected the confused thinking of sixteenth century writers and historians. And who confused these sixteenth century scholars? Henry Harrisse (1829-1910), the French historian, says it was mainly Sebastian Cabot, "a man capable of disguising the truth, when it was to his interest to do so." [5] Here then is Stephen Leacock's commentary on John Cabot's second voyage:

> The second expedition sailed from the port of Bristol in May of 1498. John Cabot and his son Sebastian were in command; of the younger brothers we hear no more. But the high hopes of the voyagers were doomed to disappointment. On arriving at the coast of America Cabot's ships seem first to have turned towards the north. The fatal idea, that the empire of Asia might be reached through the northern seas already asserted its sway. The search for a north-west passage, that will-o'-the-wisp of three centuries, had already begun. Many years later Sebastian Cabot related to a friend at Seville some details regarding this unfortunate attempt of his father to reach the spice islands of the East. The fleet, he said, with its three hundred men, first directed its course so far to the north that, even in the month of July, monstrous heaps of ice were found floating on the sea. 'There was,' so Sebastian told his friend, 'in a manner, continual daylight.' The forbidding aspect of the coast, the bitter cold of the northern seas, and the boundless extent of the silent drifting ice, chilled the hopes of the explorers. They turned

towards the south. Day after day, week after week, they skirted the coast of North America. If we may believe Sebastian's friend, they reached a point as far south as Gibraltar in Europe. No more was there ice. The cold of Labrador changed to soft breezes from the sanded coast of Carolina and from the mild waters of the Gulf Stream. But of the fabled empires of Cathay and Cipango, and the 'towns and castles' over which the Great Admiral was to have dominion, they saw no trace. Reluctantly the expedition turned again towards Europe, and with its turning ends our knowledge of what happened on the voyage.

That the ships came home either as a fleet, or at least in part, we have certain proof. We know that John Cabot returned to Bristol, for the ancient accounts of the port show that he lived to draw at least one or two instalments of his pension. But the sunlight of royal favour no longer illumined his path. In the annals of English history the name of John Cabot is never found again.[6]

Most of what Leacock wrote about the 1498 voyage is historically incorrect. Sebastian Cabot was not on the second voyage. Cabot did not go north. What Leacock is doing is confusing the voyage of 1498 with a voyage made by Sebastian ten years later.[7] None of Cabot's ships from the second voyage made it back to England, at least, there is no evidence to that effect. Yet, Leacock says, "That the ships came home either as a fleet, or at least in part, we have certain proof." Where is the "certain" proof? Surely the references in certain documents to Cabot's pension cannot be taken as evidence that he returned. His pension was paid out for many months after he left Bristol in May 1498, but in his absence this could have been paid to his wife. Leacock says nothing about the possibility that Cabot might have died somewhere on the coast of North America in 1498 or 1499, and there is no reference to the likely possibility that Cabot went so far south that he encountered the Spaniards and this encounter terminated his search for Cipango. As we have seen, there are documents that contain information that makes this interpretation very likely. These documents were all available to Stephen Leacock in 1914.

Henry K. Gibbons, a Newfoundland writer, in a book on Cabot entitled, *The Myth and Mystery of John Cabot* (1997) produced his book by using the secondary sources listed in his bibliography, even though he did reproduce several documents. Gibbons makes reference to the fact that Cabot went south and may have encountered the Spaniards, but he says nothing about what might have happened to Cabot and his men on the coast of South America. He does, however, make reference to the matter of Cabot's returning home to

Bristol, even though there is no evidence to that effect. Here is how Gibbons puts it:

> According to some sources Cabot returned to England after the voyage of 1498. The voyage was not productive, as the Far East was not attained and it had cost much time and money to the backers. Cabot was in disgrace and probably had his pension discontinued. He may have remained in Bristol a broken man or he may have returned to Genoa where his family and in-laws were living. On the other hand he may have died as some writers suggest. However, if Cabot was fit enough to lead an expedition of one hundred to two hundred men to North America and return at the age of 49 or 50 years of age, he may have started another project in Genoa or elsewhere as the records show he was not a quitter.[8]

As can be seen, Gibbons' reference to the "homecoming" is fairly close to Newton's description in 1932 and Biggar's in 1903. Obviously they all used the same sources containing the same erroneous information, noticeably the descriptions from sixteenth century secondary sources to all of which that old charlatan Sebastian Cabot, made a significant contribution.

John Cabot's story is both fascinating and frustrating because, with regard to his two voyages, he left so little information that there is no way to be certain about anything. Hence, we must make every attempt to make the best use of what information we do have. Evaluating that information in a professional manner is the work of historians, and responsible historians and writers must make every attempt to avoid unreliable reconstructions the best known of which is the one by Arthur Davies in 1955.

The American historian, Samuel E. Morison, who wrote many books about the discovery and exploration of North and South America was so frustrated with Cabot that he concluded that the whole expedition was lost in a storm on the North Atlantic and let it go at that. Despite the documents which seem to indicate that Cabot's expedition very likely ended up in the Caribbean Sea, Morison would not accept any of it.[9] If it were his theory he would probably have written a book about it, but because it was Williamson's theory, Morison completely ignored it. After Williamson's theory turned up in his books in 1929 and 1962 it almost seemed as if Morison wanted to block out that possible interpretation as to what happened to Cabot in 1498 or 1499. Here is Morison speaking for himself:

> John Cabot and his four ships disappeared without a trace. No
> report of them reached Europe. Anyone may guess whether they
> capsized and foundered in a black squall, crashed an iceberg at
> night, or piled up on a rocky coast . . . Thus, the only known facts of
> John Cabot's second voyage are that it departed Bristol in May
> 1498, that one ship returned shortly, and that Cabot and the other
> four ships were lost . . .
> The rest is silence.[10]

James A. Williamson claimed that John Cabot was not interested
in discoveries of new land and especially not interested in fish. Like
Columbus, Cabot's one objective was the riches of the East,
described by Marco Polo in his writings; Cabot was prepared to risk
his life for those riches, and in the end he died in pursuit of them.
The land that he did discover in the western ocean was to be merely
a stepping stone on his way to Cipango. Like millions before and
after him he was motivated by greed. Williamson puts it very
succinctly in these words:

> He came in therefore not only as a navigator but as a geog-
> rapher and a specialist in the spice trade and, it is clear, as the com-
> mander in full control. He would extend a fishing voyage into a new
> trade-route that would divert the richest of all trades into an
> English port. It was sound enough had the world-map been true as
> he and Columbus viewed it. But the unsuspected presence of
> America defeated them both.[11]

John Cabot's two voyages to North America are indeed a fascinating study, and a look at the bibliography in this book will give an idea
of the number of people who have written about Cabot in the last
one hundred and sixty-seven years since Richard Biddle in 1831
produced what might be called the first modern secondary source.
Part way through his research, Biddle, an American lawyer turned
historian, came to the conclusion that it was Sebastian Cabot who
was the true discoverer of North America. Hence the title of Biddle's
book, *A Memoir of Sebastian Cabot*[12]. In a way, Sebastian, who by
1831 had been dead for two hundred and seventy-four years, was still
up to his old tricks, for Biddle in his research used the secondary
sources produced by historians of the sixteenth century particularly
the historian, Peter Martyr, whose interviews with Sebastian Cabot
shaped historical thinking about the two Cabot voyages not only
during the sixteenth century but for three and a half centuries into
the future. It took historians half a century after Biddle to undo the
damage that Sebastian Cabot had done, for by then it was realized

that it was not Sebastian but John Cabot who should be credited and honoured for the voyages to North America in 1497 and 1498.[13]

Cabot's reconnaissance voyage of 1497 was a success in that he discovered a "mainland" in the western ocean which he hoped to use as a stepping-stone to the riches of Cipango. His second voyage in 1498 which was clearly an attempt to reach Asia was a failure in that nobody returned from that voyage, and hence nothing was accomplished. Robert H. Fuson, the American scholar, in the 1980's in his brilliant essay about Cabot entitled, "The John Cabot Mystique," ends his essay with these thoughts which undoubtedly leave a lasting impression in the mind of anybody studying John Cabot and his trans-Atlantic voyages in the final years of the fifteenth century:

> There will always be an aura of mystery surrounding John Cabot. That frequently happens when things are left unfinished and virtually no explanation accompanies them. That he was an exceptional man, there can be no doubt. Why else would so many of us write so much about one of whom we know so little?[14]

So, in the end, what really happened to the great Italian navigator and explorer, John Cabot, who discovered North America in 1497? He died on the way to Cipango.

Notes to Epilogue

1. Louis-Andre Vigneras, "New Light on the 1497 Cabot Voyage to America," *Hispanic American Historical Review,* Volume 36, 1956. See also Vigneras, "The Cape Breton Landfall: 1494 or 1497?" *The Canadian Historical Review,* Volume 38, September 1957.

2. Ethel S. Patterson, "Professor Arthur Davies: A Memoir" in K.J. Gregory and W.L.D. Ravenhill, (eds.), *Exeter Essays in Geography in Honour of Arthur Davies* (Exeter, Devon, England, 1971). See also Arthur Davies, "The Last Voyage of John Cabot and the Rock at Grates Cove, "*Nature,* Volume 176, November 26, 1955, pp. 996-999.

3. Henry P. Biggar, *The Voyages of the Cabots and of the Corte-Reals to North America and Greenland, 1497-1503,* (Paris, 1903), p. 81.

4. Arthur P. Newton, *The Great Age of Discovery,* (London, 1932), pp. 136-137.

5. Henry Harrisse, *John Cabot, the Discoverer of North America, and Sebastian his son*, (London, 1896, Reprint, New York, 1968), p. 115.

6. Stephen Leacock, *The Dawn of Canadian History,* (Chronicles of Canada Series, edited by G.W. Wrong and H.H. Langton), (Toronto, 1914, Reprint 1964), pp. 79-80.

7. See James A. Williamson, *The Cabot Voyages and Bristol Discovery under Henry VII* (Cambridge, England, 1962), Chapter IX.

8. Henry K. Gibbons, *The Myth and Mystery of John Cabot,* (Port aux Basques, Newfoundland, 1997), pp. 60-61.

9. Samuel E. Morison, *The European Discovery of America: The Southern Voyages, 1492-1616,* (New York, 1974), p. 191.

10. Samuel E. Morison, *The Great Explorers: The European Discovery of America,* (New York, 1978), pp. 73-74.

11. Williamson, *op. cit.*, pp. 114-115.

12. Richard Biddle, *A Memoir of Sebastian Cabot,* (London, 1831), pp. 144-152.

13. Henry Harrisse, *op.cit.*, Chapters 3 and 4.

14. Robert H. Fuson, "The John Cabot Mystique," in Stanley H. Palmer and Dennis Reinhartz (eds.), *Essays on the History of North American Discovery and Exploration* (A Walter Prescott Webb Memorial Lecture), Arlington, Texas, 1988, p. 47.

The American historian Samuel Eliot Morison, who believed that the Cabot expedition of 1498 was likely lost in a storm somewhere in the North Atlantic. Morison would not accept the theory that Cabot's expedition of 1498 was eliminated by Spanish pirates.

— Courtesy Harvard University Archives

Ten Contemporary Documents

Related to

John Cabot's Voyage of 1498

Document No. 1

News sent from London to the Duke of Milan, 24 August 1497

Note: This dispatch has been attributed to Raimondo de Soncino, Milanese ambassador in England. He, however, could not have written it, since he had landed in England only the day before. He disembarked at Dover on August 23 and was still there on the 24th (Cal. S.P., Milan, vol. 1, No. 536). From Milan Archives (Potenze estere: Inghilterra). Translation in A.B. Hinds (ed.), Calendar of State Papers, Milan, vol. 1, no. 535 (London, 1912), in Williamson (1962), pp. 208-209, and Firstbrook (1997), pp. 169-170.

'[24 August 1497] News received from England this morning by letter dated the 24th August . . . Also some months ago his Majesty sent out a Venetian, who is a very good mariner, and has good skill in discovering new islands, and he has returned safe, and has found two very large and fertile new islands. He has also discovered the Seven Cities, 400 leagues from England, on the western passage. This next spring his Majesty means to send him with fifteen or twenty ships . . .'

Raimondo de Raimondi de Soncino to the Duke of Milan, 18 December 1497

Note: From Milan Archives (Potenze Estere: Inghilterra). English translation and important parts of the original Italian in A.B. Hinds (ed.), <u>Calendar of State Papers, Milan</u>, vol. 1, no. 552 (London 1912), in Williamson (1962), pp. 209-210, Firstbrook (1997), pp. 170-171, and Parsons (1997), pp. 29-31.

Perhaps amid the numerous occupations of your Excellency, it may not weary you to hear how his Majesty here has gained a part of Asia, without a stroke of the sword. There is in this Kingdom a man of the people, Messer Zoane Cabot by name, of kindly wit and a most expert mariner. Having observed that the sovereigns first of Portugal and then of Spain had occupied unknown islands, he decided to make a similar acquisition for his Majesty. After obtaining patents that the effective ownership of what he might find should be his, though reserving the rights of the Crown, he committed himself to fortune in a little ship, with eighteen persons. He started from Bristol, a port on the west of this kingdom, passed Ireland, which is still further west, and then bore towards the north, in order to sail to the east, leaving the north on his right hand after some days. After having wandered for some time he at length arrived at the mainland, where he hoisted the royal standard, and took possession for the king there; and after taking certain tokens he returned.

This Messer Zoane, as a foreigner and a poor man, would not have obtained credence, had it not been that his companions, who are practically all English and from Bristol, testified that he spoke the truth. This Messer Zoane has the description of the world in a map, and also in a solid sphere, which he has made, and shows where he has been. In going towards the east he passed far beyond the country of the Tanais. They say that the land is excellent and temperate, and they believe that Brazil wood and silk are native there. They assert that the sea there is swarming with fish, which can be taken not only with the net, but in baskets let down with a stone, so that it sinks in the water. I have heard this Messer Zoane state so much.

These same English, his companions, say that they could bring so many fish that this kingdom would have no further need of Iceland, from which place there comes a very great quantity of fish called stockfish. But Messer Zoane has his mind set upon even greater things, because he proposes to keep along the coast from the place at which he touched, more and more towards the east, until he reaches an island which he call Cipango, situated in the equinoctial region, where he believes that all the spices of the world have their origin, as well as the jewels. He says that on previous occasions he has been to Mecca, whither spices are borne by caravans from distant countries. When he asked those who brought them what was the place of origin of these spices, they answered that they did not know, but that other caravans came with this merchandise to their homes from distant countries, and these again said that the goods had been brought to them from other remote regions. He therefore reasons that these things come from places far away from them, and so on from one to the other, always assuming that the earth is round, it follows as a matter of course that the last of all must take them in the north towards the west.

He tells all this in such a way, and makes everything so plain, that I also feel compelled to believe him. What is much more, his Majesty, who is wise and not prodigal, also gives him some credence, because he is giving him a fairly good provision, since his return, so Messer Zoane himself tells me. Before very long they say that his Majesty will equip some ships, and in addition he will give them all the malefactors, and they will go to that country and form a colony. By means of this they hope to make London a more important mart for spices than Alexandria. The leading men in this enterprise are from Bristol, and great seamen, and now they know where to go, say that the voyage will not take more than a fortnight, if they have good fortune after leaving Ireland.

I have also spoken with a Burgundian, one of Messer Zoanes companions, who corroborates everything. He wants to go back because the Admiral, which is the name they give to Messer Zoane, has given him an island. He has given another to his barber, a Genoese by birth, and both consider themselves counts, while my lord the Admiral esteems himself at least a prince.

I also believe that some poor Italian friars will go on this voyage, who have the promise of bishoprics. As I have made friends with the Admiral, I might have an archbishopric if I chose to go there, but I have reflected that the benefices which your Excellency reserves for me are safer, and I therefore beg that possession may be given of me of those which fall vacant in my absence, and the necessary steps taken so that they may not be taken away from me by others, who have the advantage of being on the spot. Meanwhile I stay on in this country, eating ten or twelve courses at each meal, and spending three hours at table twice every day, for the love of your Excellency, to whom I humbly commend myself.

London, the 18th of December, 1497

John Day to the Lord Grand Admiral

Note: This letter could have been written after Cabot's royal pension was awarded on 13 December 1497. Since mention is made of another expedition 'next year', it could have been written before the end of 1497. The medieval New Year began on 25 March, so the letter must have been written between mid-December 1497 and mid-March 1498. This document, the discovery of which is considered by historians as the most significant event in Cabotian research in the last century and a half, was discovered in Spain in 1956 by Louis-Andre Vigneras, an American historian. The English translation appeared in an article Vigneras wrote for the <u>Canadian Historical Review</u>, Volume 28, (1957), pp. 219-228. In an article published in <u>The Hispanic American Historical Review</u> in 1956, Vigneras wrote the following as a preamble to his article about the discovery of the John Day letter:

> While we were both doing research in the Archivo de Simancas last spring, Dr. Hayward Keniston knowing that I was interested in the discovery of North America, mentioned that he had seen a document relating to English voyages in one of the first three legajos of Estado de Castilla. My first search through the three legajos yielded nothing, but a second attempt three weeks later established the fact that the document in question was folio 6, legajo 2.
>
> I immediately realized its importance. It is a letter from an English merchant named John Day to the Grand Admiral of Castille, Don Fadrique Enriquez, in answer to a request for further information concerning English voyages. Although it bears no date and does not mention John Cabot by name, there is no doubt that it refers to Cabot's 1497 voyage.

Dr. Vigneras later related that the document was described on its cover as concerning an English voyage to Brazil which researchers took to mean a document about an English voyage to the South American country. To my knowledge no translation of this document other than that done by Vigneras in 1957 exists. Although this letter is about Cabot's voyage of 1497, there are references here to the voyage of 1498. Vigneras translation appears in Williamson (1962), p. 211-214, Firstbrook (1997), pp. 172-173, and Parsons (1997), pp. 32-33. From Archivo General de Simancas, estado de Castilla, Spain, leg. 2, fol. 6. The Spanish text was published by Vigneras (1956) and the English translation given here is from Vigneras (1957). Since 1957 extracts of this letter have appeared in several books and anthologies.

JOHAN DAY TO THE MOST MAGNIFICENT AND MOST WORTHY LORD THE LORD GRAND ADMIRAL

Your Lordship's servant brought me your letter. I have seen its contents and I would be most desirous and most happy to serve you. I do not find the book *Inventio Fortunata*, and I thought that I (*or* he) was bringing it with my things, and I am very sorry I cannot find it because I wanted very much to serve you. I am sending the other book of Marco Polo and a copy of the land which has been found. I do not send the map because I am not satisfied with it, for my many occupations forced me to make it in a hurry at the time of my departure; but from the said copy your Lordship will learn what you wish to know, for in it are named the capes of the mainland and the islands, and thus you will see where land was first sighted, since most of the land was discovered after turning back. Thus your Lordship will know that the cape nearest to Ireland is 1800 miles west of Dursey Head which is in Ireland, and the southernmost part of the Island of the Seven Cities is west of Bordeaux River, and your Lordship will know that he landed at only one spot of the mainland, near the place where land was first sighted, and they disembarked there with a crucifix and raised banners with the arms of the Holy Father and those of the King of England, my master; and they found tall trees of the kind masts are made, and other smaller trees and the country is very rich in grass. In that particular spot, as I told your Lordship, they found a trail that went inland, they saw a site where a fire had been made, they saw manure of animals which they thought to be farm animals, and they saw a stick half a yard long pierced at both ends, carved and painted with brazil, and by such signs they believe the land to be inhabited. Since he was with just a few people, he did not dare advance inland beyond the shooting distance of a cross-bow, and after taking in fresh water he returned to his ship. All along the coast they found many fish like those which in Iceland are dried in the open and sold in England and other countries, and these fish are called in English "stockfish"; and thus following the shore they saw two forms running on land one after the other, but they could not tell if they were human beings or animals; and it seemed to them that there were fields where they thought might also be villages, and they saw a forest whose foliage looked beautiful. They left England toward the end of May, and must have been on the way 35 days before sighting land; the wind was east north-east and the sea was calm going and coming back, except for one day when he ran into a storm two or three days before finding land; and going so far out, his compass needle failed to point north and marked two rhumbs below. They spent about one month discovering the coast and from the above mentioned cape of the mainland which is nearest to Ireland, they returned to the coast of Europe in fifteen days. They had the wind behind them, and he reached Brittany because the sailors confused him, saying that he was heading too far north. From there he came to Bristol, and he went to see the King to report to him all the above mentioned; and the King granted him an annual pension of twenty pounds sterling to sustain himself until the time comes when more will be known of this business, since with God's help it is hoped to push through plans for exploring the said land more thoroughly next year with ten or twelve vessels — because in his voyage he had only one ship of fifty "toneles" and twenty men and food for seven or eight months — and they want to carry out this new project. It is considered certain that the cape of the said land was found and discovered in the past by the men from Bristol who found "Brasil" as your Lordship well knows. It was called the Island of Brasil, and it is assumed and believed to be the mainland that the men from Bristol found.

Since your Lordship wants information relating to the first voyage, here is what happened; he went with one ship, his crew confused him, he was short of supplies and ran into bad weather, and he decided to turn back.

Magnificent Lord, as to other things pertaining to the case, I would like to serve your Lordship if I were not prevented in doing so by occupations of great importance relating to shipments and deeds for England which must be attended to at once and which keep me from serving you; but rest assured, Magnificent Lord, of my desire and natural intention to serve you, and when I find myself in other circumstances and more at leisure, I will take pains to do so; and when I get news from England about the matters referred to above — for I am sure that everything has to come to my knowledge — I will inform your Lordship of all that would not be prejudicial to the King my master. In payment for some services which I hope to render you, I beg your Lordship to kindly write me about such matters, because the favour you will thus do me will greatly stimulate my memory to serve you in all the things that may come to my knowledge. May God keep prospering your Lordship's magnificent state according to your merits. Whenever your Lordship should find it convenient, please remit the book or order it to be given to Master George.

I kiss your Lordship's hands,

Johan Day

Document No.4

Payments of John Cabots pension

(i)Payments of Cabots pension by Bristol customs, 25 March 1498
Note: From Public Record Office, Exchequer 122, 20/11 (View of Ac-
count, Bristol Customs, Michaelmas-Easter, 13 Hen. VII). Original
Latin, with translation, given by Biggar (1911), pp. 25-27, and by Wil-
liamson (1962), pp. 218-219.

[Among other items] ' . . . And 10 pounds paid by them to John
Calbot a Venetian, late of the town of Bristol aforesaid, for his an-
nuity of 20 pounds a year granted to him by our said lord the king by
his letters patent, to be taken at two terms of the year out of the customs and subsidies arising and growing in the said port of the town of
Bristol, to wit, for the term of the Annunciation of the Blessed Virgin
Mary [25 March 1497] falling within the time of this view, by a quit-
tance of the said John, shown upon this view and remaining in the
possession of the said collectors.'

(ii) Payment of Cabots pension by Bristol customs, 1498-99
Note: From Westminster Chapter Archives, Chapter Muniments, 12243
(Roll of Accounts of the Bristol Customers for the years 1496-99).
Original Latin, in facsimile, in E. Scott and A.E. Hudd (eds), The
Cabot Roll (Bristol, 1897). Translation found in Biggar (1911) and
Williamson (1962).

[Among other items, Michaelmas, 1497 Michaelmas, 1498] 'And
in the treasury in one tally in the name of John Cabot, 20 pounds.'
[Michaelmas, 1498 − Michaelmas, 1499] 'And in the treasury one
tally in the name of John Cabot, 20 pounds.'

Document No. 5

Polydore Vergil on John Cabot

Note; From <u>The Anglica Historia of Polydore Vergil</u>, edited with a translation from the original Latin by Denys Hay, Camden Series, vol. LXXIV (Royal Historical Society, 1950), pp. 116-117. The section given here is not in the printed editions but can be found in Liber XXIV of a MS copy, made in 1512-13, held in the Vatican Library. The names 'Ioanne Cabot' and 'Ioanne' were inserted later in spaces previously left blank (editors footnote, p. 116). English translation in Williamson (1962), pp. 224-225 and Firstbrook (1997), p. 176.

'There was talk at about this time that some sailors on a voyage had discovered lands lying in the British ocean, hitherto unknown. This was easily believed because the Spanish sovereigns in our time had found many unknown islands. Wherefore King Henry at the request of one John Cabot, a Venetian by birth, and a most skilful mariner, ordered to be prepared one ship, complete with crew and weapons; this he handed over to the same John to go and search for those unknown islands. John set out in this same year and sailed first to Ireland. Then he set sail towards the west. In the event he is believed to have found the new lands nowhere but on the very bottom of the ocean, to which he is thought to have descended together with his boat, the victim himself of the self-same ocean; since after that voyage he was never seen again anywhere.'

Second letters patent granted to John Cabot, 3 February 1498

Note: From Public Record Office, London, Warrants for Privy Seal, c. 82/173, 13 Hen. VII, February. The PRO also holds a Latin copy (Treaty Roll 179, membr. 1). Text given by Biggar (1911), pp. 22-24, and by Williamson (1962), pp. 226-227.

To the kinge

Pleas it your highnesse. Of your moste noble and habundaunt grace, to graunte to John Kabotto, Venician, your gracious letters patentes in due fourme to be made accordyng to the tenour hereafter ensuyng, and he shal contynually praye to God for the preservacion of your moste noble and roiall astate longe to endure.

H[en]R[icus]Rex

To all men whom thies presentis shall come, send gretyng: Knowe ye that we of our grace especiall and for dyvers causis us movyng we have geven and graunten and by thies presentes geve and graunte to our wel beloved John Kaboto, Venician, sufficiente auctorite and power that he by hym, his deputie or deputies sufficient may take at his pleasure vi [6] englisshe shippes in any porte or ports or other place within this our realme of Englond or obeisaunce, so that and if the said shippes be of the bourdeyn of cc [200] tonnes or under, with their apparaill requisite and necessarie for the saveconduct of the seid shippes, and theym convey and lede to the londe and Iles of late founde by the seid John in oure name and by our cammaundemente, paying for theym and every of theym as and if we shuld in or for our owen cause paye and noon otherwise.

And that the seid John by hym, his deputie or deputies sufficiente maye take and receyve into the seid shippes and every of theym all suche Maisters, Maryners, pages and our subiectes, as of their owen free wille woll goo and passe with hym in the same shippes to the seid londe or Iles withoute any impedymente, lett or perturbaunce of any of our officers or ministres or subiectes whatsoevir they be by theym to the seid John, his deputie or deputis an allother our seid subiectes or any of theym passing with the seid John in the seid shippes to the seid londe or Iles to be doon or suffer to be doon ar attempted. Yeving in commaundement to all and every our officers, ministres and subiectes seying or heryeng thies our letters patentes, without any ferther commaundement by us to theym or any of theym to be geven, to perfourme and socour, the seid John, his deputies and all our seid subiectes so passyng with hym according to the tenour of thies our letters patentes, any statute, acte or ordenaunce to the contrarye made or to be made in any wise notwithstanding.

Document No. 7

Agostino de Spinula to the Duke of Milan, 20 June 1498

Note: From Milan Archives (Potenze estere: Inghilterra). Translation in A.B. Hinds (ed.) Calendar of State Papers, Milan, vol, 1, no. 571 (London, 1912), in Williamson (1962), pp. 227 and Firstbrook (1997), p. 177.

'[London, 20 June 1498] . . . There were three other letters, one for Messer Piero Carmeliano, one for Messer Piero Penech, and one for Messer Giovanni Antonio de Carbonariis. I will keep the last until his return. He left recently with five ships, which his Majesty sent to discover new lands.'

Document No. 8

Pedro de Ayala to the Spanish sovereigns, 25 July 1498

Note: From Archivo General de Simancas, Estado Tratados con Inglaterra, Spain, leg. 2, fol. 196. The original was sent to Spain with most of the words in cipher and on receipt was deciphered inaccurately. The document was redeciphered by G.A. Bergenroth, a Spanish historian, in Calendar of State Papers, Spain (1862) and again, more accurately, by Biggar (1911), pp. 27-29; also given in Williamson (1962), pp. 228-229 and Firstbrook (1997), p. 178.

[London, 25 July 1498] . . . I think Your Highnesses have already heard how the king of England has equipped a fleet to explore certain islands or mainland which he has been assured certain persons who set out last year from Bristol in search of the same have discovered. I have seen the map made by the discoverer, who is another Genoese like Columbus, who has been in Seville and at Lisbon seeking to obtain persons to aid him in this discovery. For the last seven

83

years the people of Bristol have equipped two, three [and] four caravels to go in search of the island of Brazil and the Seven Cities according to the fancy of this Genoese. The king made up his mind to send thither, because last year sure proof was brought him they had found land. The fleet he prepared, which consisted of five vessels, was provisioned for a year. News has come that one of these, in which sailed another Friar Buil, has made land in Ireland in a great storm with the ship badly damaged. The Genoese kept on his way. Having seen the course they are steering and the length of the voyage, I find that what they have discovered or are in search of is possessed by Your Highnesses because it is at the cape which fell to Your Highnesses by the convention with Portugal. It is hoped they will be back by September. I let [?will let] Your Highnesses know about it. The king has spoken to me several times on the subject. He hopes the affair may turn out profitable. I believe the distance is not 400 leagues. I told him that I believed the islands were those found by Your Highnesses, and although I gave him the main reason, he would not have it. Since I believe Your Highnesses will already have notice of all this and also of the chart or mappermonde which this man has made, I do not send it now, although it is here, and so far as I can see exceedingly false, in order to make believe that these are not part of the said islands . . .

Document No. 9

Extracts from the patent granted by the Spanish sovereigns to Alonso de Hojeda (Ojeda), 8 June 1501.

Note: From Archivo General de Simancas, Cedulas, no. 5. Original Spanish given by Martin Fernandez de Navarrete, Coleccion de los viages y descubrimientos, vol. III (Madrid, 1829), pp. 85-88. English translation in Williamson (1962), pp. 233-234 and Firstbrook (1997), p.180. I have made minor changes in the translation.

Licence to Ojeda to pursue his discoveries on terms including the following:

Firstly, that you may not touch in the land of the pearl-gathering, of that part of Paria from the coast of the Frailes and the gulf this side of Margarita, and on the other side as far as Farallon, and all that land which is called Citrians, in which you have no right to touch.

Item: that you go and follow that coast which you have discovered, which runs east and west, as it appears, because it goes towards the region where it has been learned that the English were making discoveries; and that you go setting up marks with the arms of their Majesties, or with other signs that may be known, such as shall seem good to you, in order that it be known that you have discovered that land, so that you may stop the exploration of the English in that direction.

Item: that you the said Alonso de Hojeda, for the service of their Majesties, enter that island and the others that are around it which are called Quiquevacoa in the region of the main land, where the green stones are, of which you have brought a sample, and that you obtain as many as you can, and in like manner see to the other things which you brought as specimens in that voyage.

Item: that you the said Alonso de Hojeda take steps to find out that which you have said you have learned of another gathering place of pearls, provided that it be not within the limits above mentioned, and that in the same way you look for the gold-mines of whose existence you say you have news. . . .

And their Majesties, in consideration of what you have spent and the service you have done, and are now bound to do, make you the gift of the governorship of the island of Caquevacoa, which you have discovered, during their pleasure . . . Likewise their Majesties make you gift in the island of Hispaniola of six leagues of land with its boundary, in the southern district which is called Maquana, that you may cultivate it and improve it, for what you shall discover on the coast of the main land for the stopping of the English, and the said six leagues of land shall be yours for ever.

Document No. 10

Pietro Pasqualigo, Venetian Ambassador in Portugal, to his brothers in Venice, 19 October 1501

Note: From Paesi nouamente retrouati (Vicenza, 1507) lib. VI, cap. CXXVII. English translation from Williamson (1962), pp. 229-230 and Firstbrook (1997), pp. 178-179.

[Lisbon, 19 October 1501] On the eighth of the present month arrived here one of the two caravels which this most August monarch sent out in the year past under Captain Gaspar Corterat to discover land towards the north; and they report that they have found land two thousand miles from here, between the north and the west, which never before was known to anyone. They examined the coast of the same for perhaps six hundred to seven hundred miles and never found the end, which leads them to think it a mainland. This continues to another land which was discovered last year in the north. The caravels were not able to arrive there on account of the sea being frozen and the great quantity of snow. They are led to this same opinion from the considerable number of very large rivers which they found there, for certainly no island could ever have so many nor such large ones. They say that this country is very populous and the houses of the inhabitants of long strips of wood covered over with the skins of fish. They have brought back here seven natives, men and women and children, and in the other caravel, which is expected from hour to hour are coming fifty others. These resemble gypsies in colour, features, stature and aspect; are clothed in the skins of various animals, but chiefly of otters. In summer they turn the hair outside and in winter the opposite way. And these skins are not sewn together in any way nor tanned, but just as they are taken from the animals; they wear them over their shoulders and arms. And their privy parts are fastened with cords made of very strong sinews of fish, so that they look like wild men. They are very shy and gentle, but well formed in arms and legs and shoulders beyond description. They have their faces marked like those of the Indians, some with six, some with eight, some with less marks. They speak, but are not understood by anyone. Though I believe that they have been spoken to in every possible language. In their land there is no iron, but they make knives out of stones and in like manner the points of their arrows. And yet these men have brought from there a piece of broken gilt sword which certainly seems to have been made in Italy. One of the boys was wearing in his ears two silver rings which without doubt seem to have been made in Venice, which makes me think it to be mainland, because it is not likely that ships would have gone there without their having been heard of. They have great quantity of salmon, herring, cod and similar fish. They have also great store of wood and above all of pines for making masts and yards of ships. On this account his Majesty here intends to draw great advantage from the said land, as well by the wood for ships, of which they are in want, as by the men, who will be excellent for labour and the best slaves that have hitherto been obtained. This has seemed to me worthy to be notified to you, and if anything more is learned by the arrival of the captains caravel, I shall likewise let you know.

Twenty Appendices

Related to

John Cabot's Voyage of 1498

Appendix 1

Did Cabot Reach America a Second Time?

Ian Wilson

Note: This item by Ian Wilson is found on page 46 of his book entitled John Cabot and the Matthew published in Bristol and St. John's in 1996. This book is a sort of condensed version of Wilson's excellent book entitled, The Columbus Myth published in 1991. In this book Wilson supports James A. Williamson's theory that Cabot went south and encountered the Spanish in the Caribbean. Wilson in this 1991 book and his condensed version of 1996 does an excellent job of developing the theory in detail while at the same time relating it to the contemporary documents particularly Pedro de Ayala's letter of July 25, 1498 and Hojeda's licence granted by the Spanish monarchs and dated June 8, 1501. Wilson has developed a unique style of writing; he proposes questions and then proceeds to answer his questions in detail. With the exception of James A. Williamson's book (1962), Wilson's The Columbus Myth (1991) is the best book available dealing with the theory that Cabot and his crew may have been murdered by the Spaniards. Up to the day he died Samuel E. Morison (1887-1976), the American historian, would not give this theory a second thought. (Used by permission of the publisher).

Again the Spanish are our chief source for the information that serious trouble befell Cabot's five-ship expedition. Reporting from London, envoy Pedro de Ayala told his royal masters in Spain:

> News has come in that one of these [Cabot's ships] . . . has made land in Ireland in a great storm, with the ship badly damaged. The Genoese [i.e. Cabot] kept on his way.

The fact that Cabot's 20 pound pension from Henry VII is recorded as being paid from Bristol customs receipts for the year 29 September, 1497 to 19 September, 1498, and again for the same period 1498-9, was once regarded as evidence that he did manage to return. Today, however, the prevailing historical opinion is against this. We know that the expedition had been provisioned with sufficient supplies to last them for at least a year from the time of their departure in May 1498. So it was clearly intended and expected that they would be away for a long time, inevitably, to gain the maximum benefit from the venture. Furthermore it would have been perfectly legitimate for Cabot's wife Mattea or another senior member of the family to claim the money on his behalf while they believed him still alive somewhere overseas.

What seems to indicate reasonably conclusively that Cabot never did return is a reference to him in a manuscript copy of the *Anglica Historia* written by the Italian chronicler Polydore Vergil in 1512-13. Although by this stage Vergil could not even remember Cabot's name, simply leaving a space for it to be filled in later, he remarked of what can only have been Cabot:

> . . . he is believed to have found the new lands nowhere but on the very bottom of the ocean, to which he is thought to have descended together with his boat, the victim himself of that self-same ocean; since after that voyage he was never seen again anywhere.

Even so, could all four unaccounted-for vessels, and Cabot along with them, simply have sunk without a trace somewhere in mid-ocean as a result of the 'great storm' described by Ayala? Although this was certainly the opinion of United Sates Columbus expert the late Admiral Samuel Morison, British naval historian Dr. James Williamson has disagreed:

> In the history of Atlantic exploration for the ensuring century, beginning with the Corte Reals of Portugal in 1500 and going forward to Gilbert and Frobisher and Davis and the Virginian pioneers of Raleigh's time, there is no instance of a multi-ship expedition having been entirely wiped out by an unknown disaster; and we are entitled to say that the odds were heavily against it in 1498.

But if this was the case, if one or more of the four vessels which pressed on across the Atlantic actually reached the other side, what *then* happened to them? Are there any clues that members of Cabot's 1498 expedition did arrive in America? And if so why did they seemingly never manage to make the return journey?

The English Rediscovered North America

David B. Quinn

Note: This extract by David B. Quinn is part of an Address delivered at the Annual Meeting of the Associates of the John Carter Brown Library at Brown University, Providence, Rhode Island, U.S.A. on May 14, 1964 (Published by Brown University Press in 1965). Quinn in this extract is concerned with the location of "Brasil" which he says elsewhere is undoubtedly the island of Newfoundland. Quinn mentions Cabot's "failure" in his first voyage (1497) and "success" in his second (1498). The question that begs an answer here is this one: What was accomplished by Cabot's ill-fated expedition of 1498? Notice that Quinn believes that one or more of Cabot's ships might have returned to Bristol. There is really no documented evidence to that effect, and D.B. Quinn, one of the greatest Cabotian scholars of this century, is well aware of that. (Used with permission of Brown University).

We are now on reasonably firm ground when we say that the English rediscovered North America after the Norsemen and before Columbus voyages took place. When the Englishman John Day mentioned to Christopher Columbus late in 1497 "the land which the English discovered in times past as your Lordship well knows," the tone of that tantalizingly brief reference was none the less authoritative. My hypothesis on the "Brasil" referred to by Day and other contemporaries is no more than a guess, but it is one which has a certain plausibility. This is that the land found was regarded as of little importance in itself but only as marking the location of a major offshore fishery in western waters. There are indications, if not entirely explicit evidence, that the Newfoundland Banks were first found by the English and were being systematically exploited by them at least from 1490 onwards.

How, then, about John Cabot? We shall probably have to discard his priority in the rediscovery of North America if what I have said is likely to be correct. The clever Italian then appears as one who had learned in Spain in 1493 of Columbuss discovery and who came to England as soon as possible because he had an inkling that the English too had found something in the West, only farther north and not so far distant from Europe. He consequently set out in 1495 or 1496 to use the English discovery as a jumping-off point for the protruding northeast horn of Asia as it appeared in pre-Columbian world maps of the time. Cabot's failure in his first voyage was crowned by apparent success in his second. In 1497 he sailed with full authority to occupy parts of Asia not hitherto touched by the Spaniards in the name of the King of England. He reported after his return on the fishworthiness of the Newfoundland Banks, but his main discovery, as he worked his way along some appreciable part of the coast line between Maine and Labrador, was a continental shore line he firmly believed to be Asia.

There is no direct evidence that in 1498 any one of the five ships which sailed with Cabot for Asia found herself on the barren and northwest-trending Labrador coast. Yet the chances are that one or more of them did so and that she (or they) returned to tell the tale. This, if it happened, is likely to have provided the clue that the obstinately un-Asiatic and uncivilized shores they found were not Asia at all but those of a new land of continental proportions, one which might perhaps be circumvented by an assiduous following of this northerly shore. John Cabot, it seems, was lost on the voyage; a ship equipped by King Henry returned. This is all we have. We can do more than guess, even if we cannot quite prove, that these voyages in 1497 and 1498 established with reasonable certainty that the "Brasil" which the English found in times past was no isolated island but the outlier of a continent, not Asia but another. Columbus, blindly or obsessedly, was to miss the continentality of America; Vespucci was to see it later and to his own enduring credit. John Cabot had, it is probable, seen it before his death, or at least the knowledge of it came home with his surviving ships.

Cabot Thought he had Reached Asia

William P. Cumming

*Note: This item on John Cabot comes from the book <u>The Discovery of</u>
<u>North America</u> (New York, 1971) edited by William P. Cumming,
Raleigh A. Skelton and David B. Quinn. Since Cumming wrote the first
one-third of the book, this passage is attributed to him. Notice that Cumming believed that in 1497 Cabot, "sighted land somewhere along the
Maine coast," and that on his second voyage he may have explored "the
coast to Florida or beyond . . . ." Note also the reference to La Cosa.*

In 1497 Cabot thought he had reached Asia, 'the country of the
great Khan' or the 'Island of the Seven Cities.' After making landfall
on 24 June on a wooded shore near his first sight of land, he sailed
along the coast for a month or more without going ashore before
returning to England. Where he landed is uncertain; it may have
been as far north as Labrador or near Cape Bauld, Newfoundland.
Probably, but not certainly, he sighted land somewhere along the
Maine coast; his last view of the shore was a cape, possibly Cape
Breton or Cape Race, Newfoundland.

In May 1498 Cabot set out again from Bristol in a small fleet of
four ships equipped by the merchants of that seaport and a fifth
fitted out by the king. Cabot never returned; he found, said a contemporary, 'new lands nowhere but on the very bottom of the ocean.'
No record of the voyage survives nor of how many of the ships
returned eventually to Bristol. Unsupported English claims that
Cabot or surviving ships in his fleet explored the coast to Florida or
beyond date from the latter half of the next century.

Cabot's maps, mentioned by Soncino, and his report to Henry VII
upon his return from his first voyage, are lost; but copies of his maps
reached Spain by 1498 and are reflected in a map made by Juan de la
Cosa in the summer of 1500. La Cosa (not the man of the same name
who was the owner and master of the *Santa Maria* on Columbus first
voyage) served as a seaman in the *Nina* on Columbus second voyage
of 1493.

Reports of Cabot's voyages reached Portugal as well as Spain;
several expeditions by Portuguese soon sailed for northern waters
with the encouragement of the King, Dom Manoel. The King was
concerned because the land might be within the sphere of influence,
east of the longitudinal line agreed upon and assigned to him by the
Treaty of Tordesillas in 1494, and because in that region he might
find a shorter way to the East than the route by the Cape of Good

Hope taken by Vaso da Gama, who had returned from his two year voyage in 1499.

William A. Munn, the Harbour Grace historian, who in 1905 proved that the writings or markings on the so-called "Cabot Rock" at Grates Cove had nothing to do with John Cabot's voyage of 1497 or the Cabot expedition of 1498.
    — Courtesy, the Centre for Newfoundland Studies,
    Memorial University, St. John's, Newfoundland

Appendix 4

Cabot Died Here

Mike Flynn

Note: The following Evening Telegram news item of February 8, 1995 was written by Mike Flynn, a correspondent for the St. John's newspaper. Date-lined Grates Cove, it is effectively an interview between Mike Flynn and Fred Cramm, the chief proponent of the "Cabot Rock" theory or legend in Newfoundland. Apparently Cramm was greatly impressed by Arthur Davies article connecting John Cabot and Grates Cove. Cramm places a lot of importance on "two large plaques" he saw on a visit to Bristol in 1994. Cramms family name is spelled with one "m" on the cover of his recent book. The Matthew was not on Cabot's ill-fated expedition of 1498. According to the Bristol Customs Records in the Public Record Office in Bristol (E. 122/199, 1. Account, 19-20 Henry VII), the Matthew was used in the British coastal trade; as well she made several trips to France and Spain, and Ireland between 1503 and 1505. (See Williamson 1962, p. 206). As is done quite frequently elsewhere Arthur Davies name is spelled incorrectly as "Davis." Arthur Davies (b. 1906) is a geographer not a historian.

Some people may dispute John Cabot's landing site in the New World, but there should be no question about his final resting place, says a Trinity Bay man.

Old Perlican resident Fred Cramm claims the Italian explorer died at or near Grates Cove, Trinity Bay, and he says there is sufficient historical documentation to prove the claim.

Cramm says Grates Cove should take its rightful place in Newfoundland history books. He has organized a campaign to have the area visited by the Matthew, a replica of Cabot's ship being built in Bristol, England.

The Matthew is scheduled to visit Newfoundland in 1997 in conjunction with the province's 500th anniversary celebrations. And Cramm wants to have the vessel in port during a wreath-laying ceremony at Grates Cove to honor the man credited with discovering Newfoundland in 1497.

Cramm visited Bristol three months ago and toured an exhibition near where the Matthew is being built. Of particular interest to Cramm, besides the numerous artifacts from Cabot's era, were two large plaques.

One explains why Bonavista is accepted by the British as Cabot's landing site, while the second explains why they believe Cabot spent his final days in the Grates Cove area.

"I was intrigued by this, but at the time I wasn't aware of a comprehensive article in the Nov 26, 1955, issue of *Nature* magazine written by Prof. Arthur Davis of the University of Exeter in England.

"Davis, an historian, did a tremendous amount of research and he relates how Cabot, accompanied by a second vessel, made another voyage to Newfoundland in 1498. The Matthew, however, never returned to England and it was determined that the vessel was lost near Grates Cove."

Davis concluded Cabot was lost near Grates Cove. His article also stated Portuguese explorer Gaspar Corte Real visited the Grates Cove area in 1500 and made contact with a Beothuck tribe that displayed an Italian sword and earrings believed to have belonged to Cabot and his crew.

According to Davis, accounts of Cabot's voyage brought back to England by the crew of the second vessel indicate the Matthew was wrecked near the Trinity Bay community, but Cabot and his men were believed to have reached shore.

They are said to have etched their names on a large flat rock to catch the attention of the second crew if a search was organized. Many people in the community remember seeing what has become known as Cabot Rock bearing the name Giovanni Caboto (Italian for John Cabot) as well as others.

But the rock was removed from the community in the mid-1960's by two men claiming to represent Memorial University. Its current whereabouts are unknown.

Cramm says he can't understand how the British have readily accepted Grates Cove as Cabot's final resting place while Newfoundlanders haven't. Part of the blame, he said, lies with the fact that until 1955 everyone accepted Prowse's History of Newfoundland as gospel.

Prowse, who wrote his book in 1895, claimed that nobody knew what became of John Cabot. Cramm says that's not the case anymore and that a monument in Cabot's memory should be erected at Grates Cove.

Appendix 5

The Grates Cove Stone

Lemuel W. Janes

Note: This item is from The Newfoundland Quarterly Volume LIV, Number 4, December 1955, p. 48. It is an editorial likely written by the editor at the time, Lemuel W. Janes (1889-1968), and is a reaction to the discussion surrounding Arthur Davies' article in Nature magazine, November 26, 1955. Janes reproduced Cormack's statement from The Morning Chronicle for 1873. A discerning reader will notice that there are two significant errors in Cormack's passage. Janes also recorded once again the engravings that William A. Munn (1864-1940), in 1905, claimed were on the rock. It is believed that Munn used "a flour paste" to help him determine the letters of the engravings. Despite Leo English's photograph the letterings have absolutely no connection with John Cabot.

In his "Narrative of a Journey Across the Island of Newfoundland" published in 1873 by the "Morning Chronicle" of St. Johns by W. E. Cormack there appears the following paragraph on page 9:

"On the promontory between Conception and Trinity Bays is the Point of Grates, and close to it Baccalao Island. The Point of Grates is the part of North America first discovered by Europeans. Sebastian Cabot landed here in 1496, and took possession of 'The Newfoundland,' which he discovered in the name of his employer, Henry VIII of England. He recorded the event by cutting an inscription, still perfectly legible, on a large block of rock that stands on the shore."

This was probably the first record in print of the existence of this historic stone and since that time no attempt has been made to preserve from the ravages of time the engraving.

The curator of the Newfoundland Museum. L.E.F. English, O.B.E., has a photograph of this stone which shows part of the inscription, chiselled markings; so badly weathered as to be almost illegible and in 1905 the late W.A. Munn, who throughout his life was a keen student of Newfoundland history, examined the Grates Cove Stone, made a drawing of its dimensions and copied what was legible or discernable. His measurements showed the stone to be about seven feet high with the top about four feet and it was carved on three sides. On one side were the letters RW, RH, 1713; IS 1669: IH, AX; on another, VN 1617, N, with a stroke through it, and 1679; HM 1670, BF and B; on the third side EB, IBHM, FC and ILR.

Mr. English says he can decipher the letters TO CAB on his photograph.

For years now this stone and its carvings have been matters for speculation and it is unfortunate that no effort was made years ago to have the stone preserved or its legend preserved from ravages of the weather. Recently, however, considerable interest has been taken in this stone and in a letter to Hon. Myles Murray, Minister of Provincial Affairs, the Keeper of Ethiography of the British Museum reported that Professor Arthur Davies of University College, Exeter, had concluded from examination of early Portuguese maps that John Cabot disappeared when his ship was wrecked on his second voyage in 1498 off the northern head of Conception Bay and he concluded from his research that the stone at Grates Cove might be regarded as an authentic record of the misadventure.

This new information will, no doubt, stimulate further investigation and the Grates Cove stone will have more than passing interest to all who are students of Newfoundland history.

Arthur Davies Story of the "Consort Ship"

Arthur Davies

Note: Arthur Davies claimed that by December 1498 only Cabot's ship and a larger ship were exploring along the coast of what is now the Avalon Peninsula of Newfoundland. Davies says that the "consort ship" brought back the news of Cabot's death to Bristol in 1499. There is no documented evidence to that effect. Davies also indicates in this extract that the Cabot expedition, "discovered the continent of North America . . . in 1498-99." Notice also the confusion between John Cabot and Antonio Gaboto. This extract is from Nature *magazine, Volume 176, November 26, 1955. Davies is a well-known British geographer, not a historian.*

The consort ship must have missed them next day but, assuming they would link up, carried south as far as Cape Spear, when it became apparent something was seriously wrong. Thereafter the crew of the consort ship set up a look-out at this 'cape of waiting' and searched the coasts to the south, and then finally in December, as a last resort, they retraced their course northward and searched Conception Bay, naming it on December 8 after three months of vigil. They may have discovered wreckage of the ship, but most probably gave up the search in December, finding no trace of Cabot. The date, December 8, points clearly to the Cabot voyage of 1498, which did not return to Bristol until after March 25, 1499. The Corte Real voyages were back in Lisbon by early October on each occasion.

Resuming the voyage, the consort ship rounded C. Rase, which means 'flat cape.' English ears noted it as Cape Race, its modern name. The ship crossed the Saint Lawrence to Nova Scotia, and thereafter coasted to New England in mid-winter. The Island of Saint Johan, off Nova Scotia, indicates the feast of Saint John of Ravenna on January 12. After reaching 38^0 N. the consort ship turned back for England, where its Portuguese pilot, Juan Fernandez, the labrador or squire of Terceira in the Azores, reported the discovery of Greenland, etc., to King Henry of England and the loss of John Cabot with the King's ship. This is noted from the account of Alonso de Santa Cruz, a friend of Sebastian Cabot in Spain, writing in 1541. In his work, "The Voyages of the Cabots" (1929), J.A. Williamson quotes the following passage of Alonso de Santa Cruz: "That region of which we now wish to treat is commonly called the land of the Labrador (Greenland) . . . As for the western side toward the land of the Bacalhaos it is said that two Portuguese brothers, named the

Corte Reals, . . . asserted that the mainland, of which they reached the extreme end, was separated from that island of the Labrador by a very large and wide sea channel, of which the pilot Antonio Gaboto also had information. It was called the island of the Labrador because a Labrador of the Azores gave notice and information about it to the King of England, when he sent in search of it Antonio Gaboto, the English pilot and father of Sebastian Gaboto. . . . The land named the Bacalhaos, where the Corte Reals went to colonise, was first discovered by the pilot Antonio Gaboto, the Englishmen, by command of the King of England." It may be noted, in passing, that Williamson argues that Santa Cruz was mistaken and knew little about the matter. But Santa Cruz had accompanied Sebastian Cabot on his voyage to the Plate River in 1526 and had close connexions whereafter in Spain with him. It is probable that Sebastian Cabot was the source of this account of Santa Cruz, belatedly giving credit to his father for the discovery of the Bacalhaos before the Corte Reals went there.

These documents and maps show that the 1498 Cabot voyage discovered the continent of North America from Greenland to Delaware Bay in 1498-99. When Cabot went down with his ship, the Portuguese pilot, Juan Fernandez, completed the voyage and reported back to Henry of England. This report found its way through the Spanish Ambassador in London to the 1500 map of the north-west Atlantic. It is the only map record and the only surviving document of the last voyage of John Cabot. It serves as a complete and accurate record of that voyage and its proud achievement of discovery in the New World. Since Fernandez held his lands in Terceira under the Corte Reals, captains donatory of the island, it is almost certain that he passed on information of the Cabot discoveries to them. The Corte Real voyages of 1500 and 1501 were, therefore, the outcome of the Cabot voyages.

Appendix 7

John Cabot's Story is Brief and Disappointing

Frederick W. Rowe

Note: In these extracts Frederick W. Rowe (1912-1994) is saying that the English in the years after 1498 made no great attempt to capitalize on what John Cabot had accomplished, particularly with regard to the exploitation of the fisheries in the western ocean, and probably more importantly in colonizing the newly-discovered lands. Yet it must be remembered that King Henry VII and those who supported Cabot were essentially interested in the rich spices and jewels of Asia. After the trading ventures failed, it took the English a full century before they started to exploit the fisheries. It is interesting to note that the first official attempt in northeastern North America towards colonization — the venture headed by John Guy in 1610 — was centered on Bristol, the English port from which Cabot sailed one hundred and twelve years before. These extracts are from Dr. Rowe's book, History of Newfoundland and Labrador (1980), pp. 56-57.

Perhaps the greatest culprit in this whole community is John Cabot's son Sebastian, who has a two-fold distinction. He was recognized as a skillful mariner, perhaps the most prestigious of his age; also, the most unanimous opinion among historians and geographers was that he was an unmitigated liar. For several centuries, on his own evidence, he was accepted as the discoverer of Newfoundland and other parts of North America. His lies and contradictions served only to create confusion and frustration on matters about which there was already a paucity of reliable information . . .

Despite all the fuss, two facts are important: first, at no time did England use Cabot's voyage to justify a claim to Newfoundland; second, wherever his landfall, Cabot was a latecomer to North America, including Newfoundland. The Norsemen, and possibly others, had preceded him.

The rest of John Cabot's story is brief and disappointing. He returned home to Bristol to announce his discovery of a "New Isle," and received the acclaim of the populace and, from the King, thanks in the form of a 10 pound gift. The amount is so small, even from Henry VII, that some scholars infer a mistake in the records. He also received a more substantial expression in the form of an annual pension of 20 pounds. More important than these monetary rewards, however, was Henry's decision, early the following year, to give Cabot new letters-patent authorizing him to impress six ships. He would pay government charter rates, and set out again to the west with a view to establishing trade relations. Henry himself provided and outfitted one ship, an action that encouraged some Bristol merchants to invest "small stocks." Cabot was able to arrange for four other ships and, in May of 1498, the five ships left carrying a year's provisions. One ship developed trouble and put in at an Irish port. The other four sailed on to complete oblivion. One can only guess at their fate: foundering in a freak storm, crushed in the ice floes of Labrador or Northern Newfoundland, swept ashore in one of the pea-soup fogs for which our east coast is famous or death at the hands of hostile natives somewhere between Labrador and what is now the Eastern United States . . .

Perhaps it was the dismal fate of Cabot's second expedition, perhaps the wealth being generated by the tropical voyages of Columbus, perhaps the excitement of da Gama's voyage to India in 1499; whatever the reason, England and western Europe generally remained apathetic about exploiting or colonizing the lands reached by Cabot. The exception to this generalization was to be found in the relative speed with which Portugal, Spain and France developed the Newfoundland cod fishery and, in the case of the Spanish Basques, the whale fishery as well.

Appendix 8

The "Cabot Rock" at Grates Cove

Allan R. Guy

Note: Ever since William E. Cormack wrote about the Grates Cove rock back in the early part of the nineteenth century, historians have visited and written about the rock which no longer exists at least, it is not in Grates Cove. Probably more lies than truth have been told about the so-called "Cabot Rock." This item by Allan R. Guy was published in the Atlantic Guardian *in April 1956, Volume 13, Number 4. What Guy found at Grates Cove had nothing to do with John Cabot. Other historians tell a similar story.*

When John Cabot landed over here in June 1497 he raised the banners of England, made certain marks and returned. This is the resume of a letter written in England on December 18th of the same year. No one knows where the banners were raised nor where those marks were made, but at Grates Cove there is an inscribed rock which has been attracting attention for more than a hundred years.

It was first brought to light by W.E. Cormack who was the first to hike across Newfoundland in 1822. The writer of a newspaper account in 1860 claimed that IO CABOTO — presumably the Italian for John Cabot — was carved in the rock. Since then a number of historians and other interested persons, including the late W.A. Munn in 1905, have visited the curious monument but haven't seen any indication that Cabot landed there.

Recently Professor G.O. Rothney of Memorial University received a letter from Professor A.A. Davies of Exeter. Professor Davies stated that he has studied some old Portuguese maps, had come to the conclusion that John Cabot was lost off the Northern head of Conception Bay on his second voyage in 1498, and that the rock at Grates Cove might be regarded as original.

I first became interested in this remarkable stone about three years ago when I read an interesting little book called "Strange Facts About Newfoundland," and last year I went to see it. A few weeks ago I returned there again to take some photographs.

We drove to the end of the pavement at Carbonear, and then more slowly proceeded over the winding road to Old Perlican, itself one of the oldest fishing settlements in North America. From there we branched off on a narrow road and completed the remainder of the 115 mile drive from St. John's on a very straight road where the Newfoundland Railway once ran.

To get the pictures it was necessary to climb down over a low cliff on a rickety wooden ladder to the flat precambrian rocks near the sea. The inscribed stone is a rectangular outcrop of lava about the height of a man and has the very appearance of a natural monument. I examined the rock slowly and found that there were inscriptions on two of its sides and on top, but neither John Cabot's name nor his initials appear anywhere on the rock. The deeper inscriptions are IS1669, RW, RH1713, IS, C, S, JM, and VN1613 or 18. The date 1671 occurs faintly in a number of places, and on the front there is a simple engraving of a house with a flag on the roof.

The rock seems to be weathering fairly rapidly, and small slices or slabs have fallen from it. There is a wide crack through it which indicates that one whole side may fall and break into pieces in a few years.

Perhaps the discoverer's name was there, and perhaps that is why seamen who came later chiselled their own initials there, but if they were there, either they must have been marked faintly or were in the parts of the rock that have already fallen and crumbled. At any rate they can't be seen there now.

The rugged coastline of Newfoundland not far from Grates Cove where Arthur Davies believed Cabot was shipwrecked during the winter of 1498.

— The Author's Collection

Appendix 9

Cabot builded better than he knew

John S. Sewall

Note: The following is an extract from an article entitled, "The Value and Significance of Cabot's Discovery to the World." This item is taken from the Proceedings of the Maine Historical Society, Portland, Maine, Volume 8, 1897, pages 427-438.

Cabot builded better than he knew. His dream was India, the land of spices and gems, the Eldorado of the East, opulent with all precious things. He was ready to brave the mystery of unknown seas to reach that golden strand. What he found instead was an uncouth continent, stretching its rugged length from pole to pole, its lonely shores hoarsely echoing the surges of the great Atlantic, its hills and valleys shrouded in somber woods, its denizens wild beasts and savage men. No populous cities, no marts of trade, no bustling harbors alive with swarming fleets, no costly argosies passing and repassing along the paths of the seas.

It was a bitter disappointment. To search for the rich empires of the Orient, and stumble upon an infinite wilderness! Well might it check the ardor of the most romantic adventurer. And had not the young mariner been well endowed with the enterprise of the age and the bold spirit of his profession, he would doubtless have given over the quest and concerned himself no longer with such ungracious results. But whatever his personal emotions, the discovery was made. The secret of the new world was out. Europe learned of the great land lying fallow under western skies, waiting for the beginnings of a career. What that career might be must depend on what races should gain possession, and whether their old-world civilization could be adjusted to the crude conditions of the new.

Imagine for a moment that Cabot had somehow found a western route to the Orient; that the oceans had opened their gates wide to his progress, and no barrier had blocked his way to the treasures of the East. Who can doubt the sordid effect? What a crusade of avarice would have started from every European port. What squadrons of caravels and galleons would have plowed their way through storm and night in the headlong race to gather the glittering spoils. Look at poor Spain, as a forlorn object-lesson, set to teach the nations the debasing and debilitating effect of this mode of using the discoveries and opportunities of the world.

But this was not to be. Europe was spared the temptation. Instead, she was confronted with a problem; a problem more perplexing than ever history had yet propounded to the race. Cabot had found and laid at her feet an unused continent; what could she do with it?

... Plainly Europe had her hands full, and was in no condition to enlist in any such enterprise as taming a wilderness and planting a nation. She was not sufficiently civilized herself. And the experiment at such a time and with such materials would have been disastrous.

Ian Wilson, a modern scholar, who agrees with the theories of Williamson and Skelton that Cabot's expedition of 1498 was eliminated by the Spanish on the northern coast of South America.
— Courtesy of Simon and Schuster Limited, London, England

John Cabot: A Biographical Sketch

Raleigh A. Skelton

Note: The following essay written over thirty years ago is taken from the 1996 edition of the Encyclopedia Americana. This essay by R.A. Skelton (1906-1970), one of the greatest cartographers of the twentieth century, was written while he was Superintendent of the Map Room of the British Museum in London. "A mainland in the west" could have been Labrador. Skelton believed that in all probability Cabot landed in Maine or southern Nova Scotia, and then coasted "eastward" to Cape Race. Raleigh Ashlin Skelton died on December 7, 1970 from injuries received in a road accident.

A merchant and citizen of Venice, John Cabot (Italian, *Giovanni Caboto*) had by his own later account, been engaged in the spice trade with the Levant. This experience, and his reading of Marco Polo's description of the Far East, probably led him to formulate his project for a west-ward voyage to "the Indies." Cabot went to England with his sons in or before 1495 with a plan for sailing westward to Cathay by a more northerly, and therefore shorter, route than the route through the trade-wind zone that Columbus had followed across the Atlantic Ocean during his voyage of 1492-1493.

For about 15 years Bristol seamen had made regular voyages into the western ocean in search of new fishing grounds and of the "Island of Brasil" shown in contemporary maps, and at some date before 1494 they had discovered a mainland in the west. It was in Bristol therefore that Cabot looked for a ship and a crew. Letters patent from King Henry VII dated March 5, 1496, authorized Cabot and his sons to discover and possess lands "unknown to all Christians." In the same year Cabot set out with one ship from Bristol, but was forced to turn back.

In May 1497 he sailed again from Bristol in the ship *Matthew* (named for his Venetian wife, Mattea). After a run of 35 days he made land on June 24, probably in Maine or southern Nova Scotia. Here he went ashore and made a formal act of possession. From this point he turned back and coasted "eastward" for 300 leagues, apparently to Cape Race in Newfoundland, whence he made a fast crossing of 15 days and was back in Bristol early in August.

Although the Bristol seamen were excited by the wealth of fish observed off Newfoundland, Cabot was preoccupied with greater things. His reconnaissance had convinced him that the land he had found was "a part of Asia" or "the country of the Great Khan." This claim was accepted at the English court. A Venetian in London wrote home that Cabot was "called the Great Admiral . . . and these English run after him like mad." Henry VII granted him a gift of 10 pounds and an annual pension of 20 pounds.

In December 1497 Cabot presented to the King his proposals for a second voyage. From his original landfall he intended to follow the coast to the southwest until he came to the realm of the Great Khan, in East Asia — the source (according to Marco Polo) of "all the spices of the world." In February 1498 he received royal letters patent allowing him to impress ships and to recruit crews. Early in May Cabot sailed with five ships.

It appears that Cabot himself perished on this voyage, although one or more of his ships may have come back. If he or his companions followed his plan, they found neither Cathay nor a westerly sea-passage to East Asia. In this sense, they may have made (in the words of the historian James A. Williamson) "the intellectual discovery of America," for subsequent English explorers did not confuse America with Asia.

THE TRUE HISTORY OF GIOVANI CABOTO, GRATES COVE, AND THE CABOT ROCK

Barbara E. Shaw

Note: In August 1994 the people of Grates Cove and area held the First Annual Grates Cove Cabot Rock Celebration. Part of the celebration involved the unveiling of a Cabot Monument the wording on which effectively details "The Legend of the Cabot Rock." In May 1996 I was given a copy of an item on the Cabot Rock. In July I wrote to Barbara E. Shaw and she replied soon after. Here are her own words: "The piece you have was written by me and read by me at the Second Annual Grates Cove Cabot Rock Celebrations, Sunday, August 10, 1995. I read it dramatically, as a Town Crier, and wrote some words, such as Eng-e-land, as I pronounced them, in an appropriate 'olde' style." The Town Crier piece, which is well-written, is based on the information in Arthur Davies' article in Nature *magazine 1955. Ms. Shaw, a true proponent of the "Cabot Rock" theory, believes that there is a real connection between John Cabot and Grates Cove: "My personal thinking on the matter . . . is that John Cabot, or a member of his crew, carved that rock in 1497, and that it deliberately 'disappeared' in the mid 1960's, so that 'first landing' honours would not go to Grates Cove." The "Cabot Rock" was in place as late as August, 1969. It is my opinion that the rock was too big to move. It may have fallen into the deep water among the other rocks in the cove.*

Hear ye, hear ye, draw near one and all, and pay heed to this, the very history of Giovani Caboto, of Grates Cove Newfoundland, and of the famous Grates Cove rock.

Know ye, that in the year of our Lord 1497, under letters patent from the hand of his glorious sovereign majesty King Henry the seventh of Engeland etc., Giovani Caboto, gallant explorer and navigator, born in Genoa, did set sail from Bristol, Engeland, in his great ship, the Matthew. This Giovani Caboto did cross the mighty North Atlantic Ocean, and after perilous journey did land at what is this day named Bonavista, in our beautiful Newfoundland. He thence proceeded on to this very Grates Cove, and to Baccalieu, and then on to more southern waters. This Giovani Caboto did claim all these mighty lands for his right sovereign majesty King Henry the seventh, and for Engeland, and he did then return, across the mighty Atlantic Ocean, unto the goodly port of Bristol.

And thence, in the very following year, the year of our Lord 1498, Giovani Caboto did return unto the new found land, in his good ship the Matthew, and a consort ship, with him. Giovani Caboto did sail northward. The consort ship did sail south, and then returned to the place now called Cape Spear, named then Cape da Espera — the place of waiting, because there they did wait for Caboto to return unto them. But, Caboto and the Matthew did not return. And the consort ship, after much waiting and much searching, did return home to Engeland, alone.

What had befallen Giovani Caboto and the good ship the Matthew?

The clues are few, but revealing.

State documents housed even now in the great country of Engeland do record that John Cabot's pension, due half-yearly, was rightly paid unto his wife at Michaelmas in the year of our Lord 1498, and on March 25, 1499, but was not paid thereafter. John Cabot appeared never again in the records of Engeland. What think you of that, ye brave listeners?

It is further recorded that in the year of our Lord 1501, but three years following on the final voyage of Giovani Caboto, the explorer and adventurer Gaspar Real did visit this eastern Newfoundland, and here did take fifty Beothuk Indians prisoner. One of the natives did have in his possession a broken gilt sword, of Italian make, and one Beothuk lad was discovered to be wearing a pair of fine silver earrings, of Venice. Historians do aver that this sword had been the possession of Giovani Cabot, and the earrings the possession of his son, Sancius, then about 14 years of age. What think ye, brave listeners?

There is further a manuscript of one Plydor Virgil, written about the year 1512, Anno Domini, in which it is recorded, and I here do quote; "John Cabot . . . a skillful mariner . . . is thought to have gone down, together with his ships, being seized by the ocean itself, for after the voyage he was never seen again." End of quotation.

There is then silence, while almost three hundred long years pass.

Then, know ye all, that during the 1800's, Anno Domini, one William Cormack did record the presence at Grates Cove, near Grates Point, of a great rock upon which were to be clearly seen inscribed the names of IO CABOTO, and SANCIUS, his son. There was also seen the word SAINMALIA.

And now, during these 1900's, Anno Domini, several scholar researchers have made bold to make comment upon this very history of Grates Cove and of the Cabot Rock

One Archbishop Hawley did say that the name Grates Cove comes from the Italian gratia; thanks be to God.

One Arthur Davies, professor at the University of Exeter, who has executed much careful research on this our subject, did say that since

109

no one named SAINMALIA is known to have been connected with the expedition of Giovani Caboto, it may be that Sainmalia is the weather-worn remnant of "Santa Maria save us," or some such expression.

The research of professor Davies, into many aged maps, records, and writings does lead him to the conclusion and belief the good ship the Matthew did sink here, off Grates Point; that Cabot and his son Sancius did reach land here at Grates Cove, and that they did carve their names upon the rock of the shore, as witness and testimony for the good instruction and information of the brave crew of the consort ship, if the carvings should be found, but alas, the consort ship did not sail as far north as Grates Point, and so the carvings were not found by them.

Professor Davies does see fit to further record and aver that it cannot in any wise be the case that the Cabot Rock is a forgery. On the contrary, he records, it must be that the Cabot Rock is genuine, since no one in Cormack's time, or earlier, had seen the aged maps, writings, and documents which have only now, in these recent 1900's, Anno Domini, come to be taken notice of and studied, and, thereby, have led and caused scholars, to reconstruct these mighty, most 500 years ago, voyages of Giovani Caboto; and has caused these scholars, from their close study of these said aged maps and documents, to conclude, record, and aver, that the gallant ship the Matthew did sink down, here, on this eastern shore of Newfoundland, at this very place, Grates Cove.

And now we think on that, good people all.

Yea, verily, at this very moment, it may well be that the blue waters of these our Grates Cove shores do even now lap endlessly over the very bones of the gallant ship the Matthew.

And of Giovani Caboto himself, and of his son, Sancius, the rest is silence.

Yet know, ye all, that still, though this earth has circled this sun most 500 turns since, that verily even now, when the mighty winds do blow ashore north east over the great ocean sea, we, who dwell here in this Grates Cove, do think of Giovani Caboto, and of his walking here.

Yet where, you do ask, is this famed Grates Cove Rock of Giovani Caboto on this very day?

The tale is sad, and treacherous.

Know ye, that circa the year of our Lord 1900 and 62, persons unknown did villainously covet this famed Grates Cove Cabot Rock for their very own devises and possession, and thus, two wretched thieves did, in a media van, enter this our Grates Cove, did chisel these famed carvings of Giovani Caboto out of the very mother rock

110

of the Grates Cove shore, and did verily bear this Cabot Rock away — to place — unknown. Now what think you, gentle listeners all?

And, further, to our great grief and woe, it must sadly be reported here that many diligent various and sundry searchings and seekings have been unable to discover the present whereabouts and hiding place of this, the famed Grates Cove Rock of Giovani Caboto.

Now may it well be that this Grates Cove Cabot Rock rests even now safely, in its present place of hiding. And may it well be that it soon returns to us, unharmed. May this, our prayer, be answered.

And so, good people all, the tale is complete.

And we here, today, yea, even at this very place where in the years of our Lord 1497 and 1498, Giovani Caboto did bravely sail and set foot, and where he did verily carve his very name upon our Grates Cove shore; truly, we, gathered here, on this day, one, and all, we do celebrate, praise, rejoice, and give thanks. Gratia.

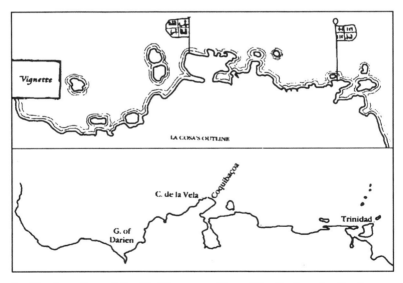

La Cosa's outline of the Caribbean coastline of South America (top) compared with the true outline indicated (bottom).
— Courtesy Simon and Schuster Limited, London, England

111

Appendix 12

The John Cabot Mystique

Robert H. Fuson

Note: The extract below comes from the book entitled, <u>Essays on the History of North American Discovery and Exploration</u> edited by Stanley H. Palmer and Dennis Reinhartz, two American historians. The six items in this book are the Walter Prescott Webb Memorial Lectures published for the University of Texas at Arlington by Texas A & M University Press, 1988. One of the lectures "The John Cabot Mystique" was written and delivered by Robert H. Fuson, formerly a professor of geography at the University of South Florida. Fuson's essay touches on the best information available about John Cabot and attempts to separate fact from fiction. Howard R. Lamar in the Introduction to the book in question says that Fuson's essay "resembles a research and historiographical spy thriller about John Cabot," who was one of the most mysterious explorers of the fifteenth century. (Used by permission of the Walter Prescott Webb Memorial Lectures Committee, the University of Texas at Arlington)

John Cabot was, and is, an enigma. For more than three hundred years he was generally regarded to have been an elderly merchant who remained in Bristol while his son, Sebastian, engaged in epic voyages of exploration. The "Sebastian supremacy," as James Williamson calls it, culminated in 1831 with the publication of Richard Biddle's *Memoir of Sebastian Cabot*. Within a few years documents were revealed that indicated that a terrible historical mistake had been made. John had made the discoveries for England, not Sebastian! By the late nineteenth century the pendulum had swung the other way: John had become the hero and Sebastian, a charlatan and weaver of fables. The "John Cabot mystique" had begun to take shape.

One hundred four years have elapsed since Henry Harrisse set out to correct the historical record and to restore John to his proper place among fifteenth-century navigators. Dozens of books and articles have appeared during the past century as scholars sorted through the old records and pored over the early nautical charts. The last significant find came as recently as 1956, when Dr. L.A. Vigneras discovered the John Day letter in the Archivo General de Simancas. Nevertheless, in spite of all of this attention to the Cabots, our ignorance is appalling.

We do not know where John Cabot was born, when he was born, or even his exact name. There were no contemporary portraits or physical descriptions that we know of, so his appearance is a mystery. No extant document informs us of his residency before he went to Venice. Only fragments of his family life have come down through the years; his father's name is in doubt, and nothing is known about his mother, two of his sons, or his brother.

No one can say when he departed Venice, and no document reveals anything about his maritime experience (if any) before or during the Venice residency. We are uninformed about what transpired after John Cabot's leaving Venice and before going to England. The year of John's arrival in England is not certain, nor are we positive about which city there was his first home. His arrival in Bristol is also undocumented.

John made a short, unsuccessful voyage in 1496, but all details of it remain a secret. There is no absolute date for the departure or return of the 1497 voyage, and the North American landing site has never been determined.

Lastly, we do not know when John died, where he died, or how he died. His death may have been at sea or anywhere on land between Canada and Florida.

Such is the stuff that makes good historical fiction, if not good history. John Cabot, unfortunately for us, was born into a world that did not keep very good records for persons of humble origin. Biographies of such men came later, after the accidents of history and geography thrust fame and fortune upon them. But John's untimely death permitted little of either. John Cabot himself left not a single holograph scrap — not a letter, not a map, not a ship's log. Nothing. Nothing, that is, except Sebastian.

The few bits and pieces of biographical information that we have are derived from certain Venetian civic records, casual references in letters mailed by contemporaries from England to correspondents in Italy and Spain, some household accounts from England, and statements made by Sebastian. Usually the latter were to people who did not know the senior Cabot personally, and Sebastian clearly kept alive memories of his father's real or imagined exploits by assigning the credits to himself.

So thoroughly did Sebastian absorb John's identity that Richard Eden, one of England's greatest sixteenth-century writers, who knew Sebastian well, was unaware that John had ever commanded a voyage for Henry VII. Further, Peter Martyr, Spain's counterpart of Eden, was also personally acquainted with Sebastian and wrote of the alleged epic voyages to discover the Northwest Passage. But Martyr never mentioned John's 1497 journey to Canada and, seemingly, had never heard of it. This is comparable to Las Casas' forgetting that there ever was a Christopher Columbus and assigning the discovery of San Salvador to Fernando!

It appears that no one in England had heard of John Cabot before 1495, no one saw him again after 1498, and almost everyone had forgotten him by 1513, the year that Polydore Vergil completed his *Anglica Historia*. Here is a man that many regard as the first post-Viking visitor to Canada and that some consider to be the true discoverer of Florida, yet the incontrovertible facts allow him only three years of English service, during which time he failed in two-thirds of the voyages he attempted.

If we sift the verifiable materials pertaining to John Cabot, we derive only a small residue of absolute truths: (1) John became a Venetian citizen between 1471 and 1473, after having resided in Venice for at least fifteen years; (2) he had at least one brother, Piero, and a Venetian wife, Mattea; (3) his father was named either Egidius or Giulio, and had been a merchant; (4) by 1484 John had at least two sons, one of whom was Sebastian; (5) between 1490 and 1493 a Venetian named John Cabot Montecalunya was in Valencia and Barcelona and may have been the historical John Cabot; (6) John was in England no later than 1495; (7) John Cabot was in London before going to Bristol; (8) on March 5, 1496, Henry VII granted letters patent to Cabot and his three sons, Lewis (Ludovico), Sebastian, and Soncio (and to their heirs and deputies) to discover and investigate lands in the eastern (i.e., East Asian), western, and northern sea; (9) John made his first English voyage from Bristol as commander in 1496, but was forced to turn back; (10) he made a second voyage in May, 1497, in the bark *Matthew*, with a crew of eighteen to twenty, including at least two Bristol merchants and two old friends; (11) a landing was made somewhere west of Ireland in June, 1497; (12) Cabot returned to Bristol in August, 1497; (13) a third voyage, with five ships, sailed from Bristol in May, 1498; (14) at least one ship from the 1498 fleet turned back from Ireland; (15) John Cabot never returned from the 1498 voyage.

These are the bare bones of John Cabot's life and his enterprise. Meat may be added to the skeleton, but only at the risk of introducing controversy. One need only turn to the exhaustive studies of the last century to ascertain the depth of the disagreement. Generally, no two Cabotian scholars have been able to come to total agreement on such matters as the number of voyages, dates of the voyages, and landfalls (among other things) because of (1) national, provincial, or personal biases, and/or (2) the role of the cartographic evidence . . .

John Cabot may have gone to England in 1494 or early 1495. He probably went first to London, then to Bristol. Here he learned that the Men of Bristol had already done what Columbus had done — failed to journey far enough. But the Men of Bristol were not looking for Asia. They many have been seeking the legendary Isle of Brazil or the Island of the Seven Cities, or perhaps nothing more than a good fishing hole. In any event, Cabot was successful in obtaining permission to sail *beyond* Brazil and the offshore islands found by the Spaniards. Once at the mainland a turn south would fetch Cathay. Also, the distance was less in these northern latitudes.

Such a northern voyage to a point beyond any European discovery would not have violated the terms of the Treaty of Tordesillas (which the good Catholic King Henry VII fully respected), it would not have been a rediscovery of something every sailor on the Bristol waterfront already knew, and it would have made Bristol the spice capital of the world.

Cabots idea was a solid one, based on a true global concept of great circle sailing. He may have gotten beyond "Brazil" in 1497, but not much beyond. His untimely death, either at sea or somewhere in eastern North America, in 1498 brought an end to the dream.

Cabot, like Columbus, never knew that he was on a new continent. Both men were dead before Europeans finally grasped the concept of a major obstruction between them and Cathay, probably around 1510 or so. But some cartographers kept the Asian idea alive until the middle of the sixteenth century.

John Cabot was quickly forgotten and his brilliant plan for a short route to Asia could not compete with the actual successes of the Spanish and Portuguese. The later attempt by the English to resurrect Cabots discovery no more impeded Cartier and the French than Drakes claim to California would slow the Spaniards in their advance. Failure to follow the 1497 voyage with colonization was the greatest geopolitical mistake the English could have made.

Appendix 13

The Rest is Silence

Samuel E. Morison

Note: Samuel E. Morison (1887-1976), the famed American historian, who spent over forty years at Harvard University, had a special interest in the early European discovery of North and South America. Morison would not accept Williamson's theory that the Cabot expedition of 1498 went as far south as the Caribbean Sea and was likely eliminated by the Spaniards. In his opinion, "John Cabot and his four ships disappeared without a trace." It was his opinion that the ship which returned to an Irish port did not complete the voyage. These extracts come from Morison's book, The Great Explorers (1978), pp. 72-75. (Used by permission).

On 3 February 1498, Henry VII issued new letters-patent for the second Cabot voyage, granting "to our well beloved John Kaboto, Venician," power to impress six English ships of 200 tuns or under, together with their tackle and necessary gear, "and theym convey and lede to the londe and Iles of late founde by the seid John in oure name," paying at the rate of government charters. Cabot also had the right to enlist any English sailors who sign willingly. Thus "Messer Zoane," says Soncino, "proposes to keep along the coast from the place at which he touched, more and more towards the East, until he reaches an island which he calls *Cipango*, situated in the equinoctial region where he thinks all the spices of the world have their origin, as well as the jewels." There he will "form a colony"; i.e. set up a trading factory, by which means London will become "a more important mart for spices than Alexandria." The king provided, manned and victualed one ship, in which divers merchants of London ventured "small stocks." Merchants of Bristol freighted four more vessels laden with "course cloth, Caps, Laces, points and other trifles," supposed to be the proper trading ruck for natives, and all five ships departed Bristol together at "the beginning of May" 1498.

As to who accompanied Master John, we are as much in the dark as on the first voyage. We know the names of but two shipmates: a Milanese cleric in London named Giovanni Antonio de Carbonariis, and a Spanish friar named Buil who had been a leading troublemaker on Columbuss second voyage. The Spanish ambassador on 25 July 1498 informed his royal master that Cabot had departed with five ships and a years provisions, and that the ship in which Father Buil sailed had put into an Irish port in distress. The other four were expected home the following month. They never returned.

Thus, the only known facts of John Cabot's second voyage are that it departed Bristol in May 1498, that one ship returned shortly, and that Cabot and the other four ships were lost. His pension was paid for the last time, probably to his wife (not yet known to be a widow) within the twelvemonth following Michaelmas 1498. Polydore Vergil, a contemporary English historian, wrote somewhat flippantly that Cabot "founde his new lands only on the ocean's bottom, to which he and his ship are thought to have sunk, since, after that voyage, he was never heard of more."

The rest is silence.

* * *

Juan de La Cosa's mappemonde dated 1500 shows a series of English royal standards planted along a coast which appears to stretch from the Labrador to Florida. A point usually identified as Cape Breton, but which may as well be Cape Bauld or Cape Race or Cape Cod, is named *Cavo de Yngleterra*. Stringing along to the westward are twelve names which make no sense and appear on no later map, and the inscription parallel to this coast reads, *mar descubierta por inglese* (sea discovered by the English). Certain historians consider this to be evidence that Cabots' second voyage ranged the coast of North America as far as Florida or even Venezuela, searching for a passage to Cathay. But, if neither he nor any of his men returned, how did Las Cosa get his facts? Obviously, since the American half of this map has been post-dated at least five years, the English flags, if anything more than a whimsy, recorded a later voyage by the Bristol-Portuguese syndicate . . .

John Cabot and his four ships disappeared without a trace. No report of them reached Europe. Anyone may guess whether they capsized and foundered in a black squall, crashed an iceberg at night, or piled up on a rocky coast. One remembers an old Irish proverb, "The waves have some mercy but the rocks have no mercy at all"; and God knows there are plenty of rocks both an and off the North American coast.

Nevertheless, John Cabot's first voyage was the herald and forerunner to the English empire in North America. Like Columbus, he never learned the significance or value of his discoveries.

Appendix 14

John Cabot's Position in the Enterprise

James A. Williamson

Note: These four extracts are taken from James A. Williamson's book, The Cabot Voyages and Bristol Discovery under Henry VII (Cambridge, England, 1962), pp. 113-115. Tryggvi J. Oleson, in his book, Early Voyages and Northern Approaches 1000-1632 (Toronto, 1963) says that these statements by Williamson constitute "a masterly summary of the achievements of one who will always occupy a leading place in the post-mediaeval history of Canada. At the same time, it unwittingly places John Cabot's son, Sebastian, in his proper niche." (p. 140)

At this stage we leave John Cabot, and may essay some estimate of him. In his Venetian period he was a merchant of enterprise sufficient to take him to Mecca and to think of new ways of working the Far Eastern trade. He had read Marco Polo and mastered the information on China and Japan which enabled him to expose the falsity of the Spanish claim to have reached Asia . . . He had studied also cartography and could express his ideas not only by maps but on a globe, a considerable professional achievement . . .

That he was an able advocate and spoke with authority is clear from his acceptance by Henry VII and the grant of the letters patent. Soncino testifies to his eloquence and conviction, and both Soncino and Pasqualigo suggest a certain magnificence and magnetism which fired the ignorant public no less than it drew support from the wary circle round the King. We may discern that John Cabot was a veritable leader of men, deprived of rounded historical greatness by premature death and lack of record. There he was unfortunate, for his son Sebastian did not play the part of Las Casa and Ferdinand Columbus, without whose devotion Christopher Columbus would not today be the great figure that he is. Sebastian did nothing for his father's reputation . . .

John Cabot's position in the enterprise may be seen more clearly since the publication of the John Day letter. The Bristol men had already found their Brasil and knew the way to it. There is no hint that they based any plan of Asiatic trade on their discovery, which was of so small promise that London, for all we can tell, had not heard of it. Cabot, seeking aid in Lisbon and Seville, where there were Bristol men to talk with, learned of it and decided that England must be his base. That would appear to have been his entry into the affair, although we lack the dates to prove it. There was land in the west, well clear of the Spanish activities. This land could be a stepping-stone on the way to Cipango. He came in therefore not only as a navigator but as a geographer and a specialist in the spice trade and, it is clear, as the commander in full control. He would extend a fishing voyage into a new trade-route that would divert the richest of all trades into an English port. It was sound enough had the world-map been true as he and Columbus viewed it. But the unsuspected presence of America defeated them both.

An aerial photo of the modern city of Bristol, showing the Avon River and St. Mary Redcliffe Church. On both expeditions Cabot left from the area of the Avon River directly opposite St. Mary Redcliffe Church.
— Courtesy J. A. Dixon, publishers, London, England

119

Appendix 15

Treasures of Spices and Precious Stones

William Macdonald

Note: What follows is an extract from an article by Macdonald entitled, "The Landfall of Cabot and the extent of his discoveries." This item is taken from the Proceedings of the Maine Historical Society, Portland, Maine, Volume 8, 1897, pp. 416-426. This writer's thinking is consistent with the theory of the French historian, Henry Harrisse, who published a book about Cabot's voyages in 1896. Note the reference to Cabot's arriving in "a golden city."

The landfall of Cabot was very probably somewhere on the coast of Labrador, and the extent of his discoveries embraced a portion of that coast and the eastern coast of Newfoundland. More than this I do not think can be safely asserted. Whether the landfall is to be located at or near Sandwich Bay, or whether Cabot followed the Labrador coast as far north as Cape Chudleigh, at the entrance of Hudson's Strait, we do not know, and I am unable to see anything in the documentary sources that can help us to find out. It would be gratifying, of course, if we could mark the spot as precisely as Plymouth Rock marks the landing-place of the Pilgrims; but we cannot do this in the case of Cabot.

It must have seemed to John Cabot an inhospitable region, this strange and new Labrador coast. No man appeared to welcome him on the morning of the twenty-fourth of June, 1497, nor did he ever know either the nature or the extent of the new world he had discovered, and little dreamed of the honor that after times would pay to his achievement. But he was the first Englishman, the first man of our race, to set foot on the North American continent; and although he did not find the treasures of spices and precious stones which his fancy pictured, he did, we may trust, on his second and final voyage, come to a golden city, though by the way of a watery grave.

Appendix 16

Discovering a Whole New Continent

Ian Wilson

Note: In this extract from Ian Wilsons popular and well-written book, The Columbus Myth *(1991), the author is exploring the possibility that Cabot's expedition or at least part of it, ended up in the Spanish sphere of influence in 1499, and encountered the Spanish desperado, Alonso de Hojeda, and his crew. Wilson also makes reference to the knowledge of America's geography that Cabot and his men would have gained had they sailed the full length of the coast of North America. This extract comes from pages 136-137. (Used by permission).*

Since we know that Hojeda and his companions reached Hispaniola in early September 1499, it would have been in August or earlier that they were in the region known as Coquibacoa, August 1499 being merely 15 months after Cabot and his men had set out from Bristol on the voyage from which they never returned. We can also be quite certain that there had been no further English voyage of transatlantic discovery, not least in view of no news having been received of Cabot. Although it had been hoped that his expedition might return to England by September 1498, it had been provisioned for a year, so there would have been no special surprise that it might take at least that long to return. Indeed, at this very time, September 1499, Vasco da Gama arrived back in Lisbon, having been away more than two years on his voyage to India.

It is therefore of extraordinary interest that in 1829 the highly authoritative Spanish historian Martin Fernndez de Navarrete, the very man whose assiduous researches brought to light Las Casass copy of Colombus Journal, wrote in the third volume of his definitive history of the Spanish voyages of discovery:

> It is certain that Hojeda in his first voyage [that of 1499] en-countered certain Englishmen in the vicinity of Coquibacoa.

'Certain Englishmen' all that way south in 1499, when supposedly so far the only English feet on American soil had been those of Cabot's few companions in the environs of Nova Scotia in June 1497? The statement is one of the most tantalisingly enigmatic in the whole history of America's discovery, all the more so because of the insistent words 'it is certain,' and because Navarrete, normally conscientious about quoting his sources, curiously omitted to do so in this instance.

121

So had Cabot and/or those who had survived from his 1498 voyage, diligently searching, like Columbus, for Marco Polo's 'Cipango . . . in the equinoctial region,' been all this time making their way steadily southwards down the eastern American coastline, only to come face to face with Hojeda and his band of Spanish desperados in Coquibacoa?

It is a tantalising thought that had they indeed done so they would have been virtually bound to have acquired more knowledge of America's geography than any other Europeans at this time, their peregrinations southwards inevitably convincing them that this had to be a whole new continent, just as Hojeda's men's coasting northwards had gained a corresponding understanding of America's southern half. Effectively they would have been the first to become aware of something approaching America's true vastness.

But it is also a much more sobering thought to realise that on meeting Hojeda any luck they might have had previously would very quickly have run out. Quite aside from the possibility of any secret orders that Fonseca may have given, Hojeda's well-practised ruthlessness towards natives and even fellow-Spaniards is indication enough that he and his men would have had little compunction about liquidating any such stray group of Englishmen, particularly since he could justify his actions on the grounds that they were trespassing on territory already allotted to Spain. Given that such an encounter ever took place, the English party may already have been enfeebled by the fevers and malaria endemic in this part of Columbia. But it is possible that they put up quite a resistance, since it was immediately after Coquibacoa that Hojeda's ships needed substantial repairs (though this may simply have been because of the tropical climates ravages upon wooden hulled vessels).

Appendix 17

Ambiguity in Some of the Early Records

Raleigh A. Skelton

Note: The following extract is taken from the essay Skelton wrote about John Cabot for the <u>Dictionary of Canadian Biography</u>, Volume I (1966). This extract is concerned with the ambiguities and confusion existing in the early records particularly the confusion between John Cabot's voyage of 1498 and Sebastian Cabot's voyage almost ten years later. The confusion originated with Sebastian, who gave Peter Martyr erroneous information. For three hundred and fifty years or more this confusion created a nightmare for historians. (Used by permission of the University of Toronto Press).

During the 16th century John Cabot's reputation was eclipsed by that of his son Sebastian, to whom the discovery of North America made by his father came to be generally attributed. This misapprehension, which Sebastian did nothing to remove before his death in 1557, arose in part from confusion between John Cabot's expedition of 1498 and the later westward voyage made by Sebastian under the English flag, in part from ambiguity in some of the early records (including those derived from Sebastian) relating to the Cabot voyages, and in part from ignorance of other records. Thus the passage in the London chronicle describing the expedition of 1498 referred to its leader simply as "a Venetian," whom John Stow, in *The chronicles of England from Brute unto this present yeare of Christ, 1580* (London, 1580) and Richard Hakluyt, in *Divers voyages touching the discoverie of America* (London, 1582), identified as Sebastian Cabot, although Hakluyt printed, also in his *Divers voyages*, the letters patent issued to John Cabot and his sons in March 1496. Until the second quarter of the 19th century historians could still credit Sebastian with the conduct of his father's two expeditions, identifying that of 1498 with the voyage into high arctic latitudes made by Sebastian and described by him to Peter Martyr between 1512 and 1515 and to Ramusio in 1551. The Spanish, Venetian, Milanese, and English documents which came to light during the 19th century enabled Henry Harrisse (1882 and 1896) to restore to John Cabot the credit for the ventures of 1497 and 1498, and G. O. Winship (1900) to review the chronology and to isolate the aims and course of Sebastian's independent voyage, which he assigned to the years 1508-9.

Although some 20th-century historians associate Sebastian's statements about his own voyage with that of 1498, Winship's views, as adopted and developed by J. A. Williamson and R. Almagia, have commanded fairly general consent. They establish indeed a coherent and progressive relationship between the geographical concepts and objectives of the three voyages. As the scanty records place beyond doubt, John Cabot sailed in 1498 with the intention of running southwest from his discovery of 1497 along the coast which he supposed to be that of East Asia. If, as seems probable, either he or his companions executed this design, they found neither Cathay nor any westward sea passage. That this "intellectual discovery of America" may have resulted directly from the 1498 voyage is suggested by the fact that (as Williamson has pointed out) the records of subsequent English voyages to the west contain "no more talk of Asia as lying on the other side of the ocean." By 1508 Sebastian Cabot was seeking a northern passage round the continent which lay across the seaway to Cathay.

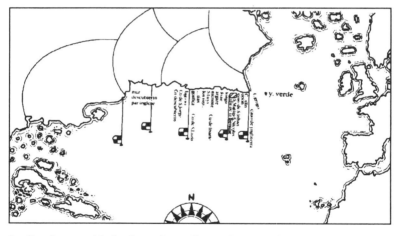

La Cosa's map with the flagged coastline attributed to English discovery.
— Courtesy Cambridge University Press
and the British Museum, London, England

Appendix 18

The Purpose of the Voyage of 1498

James A. Williamson

Note: James A. Williamson (1986-1964) who has been called "the dean of Cabotian studies" was an English scholar who was associated for many years with the Hakluyt Society. He published two important books related to Cabot's voyages, one in 1929 and the other in 1962. The extracts below are taken from the 1929 book entitled <u>The Voyages of the Cabots and the English Discovery of North America under Henry VII and Henry VIII</u> (pp. 179-180-181). Serious students of Cabot should begin by reading these two books. Williamson's main ideas about the Cabot voyages were summarized in a historical booklet published in 1937. These extracts give a somewhat detailed account of the steps leading up to Cabots expedition of 1498. As Williamson indicates this trading venture was probably more significant than Cabot's reconnaissance voyage in 1497. Notice that at this time (1929) Williamson's belief is consistent with that of Morison: "The Atlantic swallowed Cabot and all his men, and not one of them can be positively proved ever to have come home."

We have already seen that the plans for John Cabot's second voyage were being discussed immediately after his return in 1497. The first record of active preparations is comprised in the second letters patent issued to him on February 3, 1498. This document authorizes Cabot or his deputies to take six English ships, of a burden not exceeding 200 tons, for a voyage to "the land and isles" lately found by him, on payment of the rates of hire ordinarily paid by the Crown. He may also take with him all persons who will voluntarily join the venture. All officers and subjects of the king are enjoined to facilitate these arrangements, which may be carried out despite any statute or ordinance "made or to be made" to the contrary . . .

The purpose of the voyage was that expressed in the letters of 1497, to follow the new coast to tropical latitudes and thence to discover the spice regions adjoining Cipango. There is no further mention of the convicts whom it was said the king had promised Cabot for the purpose of founding a colony; and we do not know if they actually sailed or not. A project for a colony would seem to be rather out of place in opening relations with the powerful princes who were supposed to rule over the regions sought for. It is possible that the settlement was intended to be on the coast found in 1497, where it would have been useful as half-way depot at which goods from the spice islands could be collected and thence forwarded to England.

This plan for a half-way post on a long route is noticeable in more than one Elizabethan scheme for trade with Cathay. If the convicts were employed there would have been no need to mention them in the patent of 1498; their passport would have been a different kind of document.

The shipping was equipped at Bristol. According to the chronicles a large vessel was manned, equipped and victualled at the kings cost. The second patent does not promise any royal contribution, but that is not proof that the chroniclers are wrong; for it is concerned only with ships that Cabot and his deputies were to impress at their own expense, and no patent would have been necessary to enable the king to hire a ship. In addition there were three or four small ships laden by the merchants of London and Bristol. The Privy Purse entries give the names of three of these Londoners. On March 22, 1498, Lancelot Thirkill of London received a loan of 20 pounds "for his ship going towards the new Island," and a second note of the same date records the delivery of (apparently) another sum of 20 pounds to the same man, "going towards the new Isle." On April 1st Thirkill and Thomas Bradley jointly received another loan or gift of 30 pounds, whilst John Carter, "going to the New Isle," was paid forty shillings "in reward." The inference is that all three proceeded on the voyage. Another person who is certainly known to have sailed was an Italian named Giovanni Antonio de Carbonariis, who was recorded on June 20 to have "left recently with five ships which His Majesty sent to discover new islands." The Bristol merchants, as before, are unnamed, and we are left to guess that Thorne and Elyot may have been of their number. The cargoes consisted of "slight and gross merchandises, as coarse cloth, caps, laces, points and other trifles," which would have yielded a glorious profit if exchanged for an equal bulk of spices. The whole fleet, according to De Ayala, was provisioned for a year; and since he added that "they" hoped to be back by September, it may be that the provisioning was for a colony to remain in the new land.

The date of sailing is not precisely fixed. Fabyan's Chronicle says "in the beginning of May"; the anonymous Chronicle, "in the beginning of summer"; and Spinula, on June 20, "recently." We can regard May as the probable month. On July 25 Pedro de Ayala wrote some additional details. The five ships, he says, ran into a storm after leaving Bristol, and one of them put into an Irish port in a damaged condition; but Cabot with the others continued on his way. In the ship that fell out there was "another Friar Buil," an allusion to a missionary who had accompanied Columbus. Coupling this remark with Soncino's statement that some Italian friars purposed going to the new land, we may guess (it is only a guess) that the man in question was the Giovanni Antonio of Spinula's letter. De Ayala's news contains the last positive information yet discovered about John Cabot's second expedition. The chroniclers have nothing further, save that no tidings had been heard by September. From that day in May or June when the crippled ship made port the Atlantic swallowed Cabot and all his men, and not one of them can be positively proved ever to have come home.

Today the Ostrich Inn in Bristol occupies the site almost directly opposite the site on the Avon River from which Cabot departed on his expeditions in 1497 and 1498.

— The Author's Collection

Appendix 19

John Cabot and the La Cosa Map

John H. Gilchrist

Note: This extract is from an article in American Neptune, Volume 45, Number 4, 1985. The writer is John H. Gilchrist, an American naval architectural engineer, recently turned writer. The article is entitled, "Cabotian Conjectures: Did a Cabot Reach Maine in 1498?", pp. 249-250. In this article Gilchrist questions the date of the La Casa map, and is in agreement with Samuel E. Morison that the Western Hemisphere section of the map was not completed until at least 1505. Skelton, Williamson, and Quinn reject this line of reasoning. In a footnote Gilchrist says there is a good coloured reproduction of the La Cosa map in the National Geographic, November 1975. This extract gives a good indication of the way historians argue about controversial subjects particularly about important cartographical documents such as the La Cosa map. I take the position that the La Cosa map is related to Cabot's two voyages of 1497 and 1498 and that the English standards (flags) on it range from Cape Race to the southern coast of Maine, and that the "sea discovered by the English" (mar descubierta por inglese) is that area of the Atlantic Ocean south of Newfoundland and east and southeast of Nova Scotia. (Used by permission).

The period from the early nineteenth century to the first decade of the twentieth witnessed historiographic contributions to North and Central American discovery that were of the highest order. Not least was Alexander van Humboldt's identification in 1833 of the now celebrated La Cosa portolan mappemonde (ca. 1500-1505). This great map, the earliest known post-Columbian map of the New World, is no less remarkable for conceding to England, in symbol and word, that portion of the coast from Cape Race, Newfoundland to a promontory that might be Cape Elizabeth.

Why should Juan de La Cosa, a Spanish mariner and royal cartographer in an epoch of Iberian maritime dominance, have credited England with New World territory in the absence of conclusive evidence that anyone, least of all the English, had navigated below Cape Breton prior to 1520? One explanation for the English standards La Cosa staked along the northeast coast, spanned by an inscription *saying* that the coast was discovered by Englishmen, is that Bristol mariners had made pre-Columbian visits to Baccalao, as Newfoundland, whose waters won early fame for their codfish, was called, as early perhaps as 1489. Such a thesis, however, assumes voyages about whose existence there is even more doubt than that of the Cabots in 1498. A more plausible case, favored by D. B. Quinn

128

and others, is that La Cosa got his New World information from John Day's letter to Columbus, which, written in December 1497 or January 1498, necessarily describes the Cabot voyage of 1497. The trouble with this explanation of the English standards is that La Cosa appears to have documented a coastline exceeding the geographical confines of Day's description. It is true that if the latitudinal bandwidth defined by La Cosa's 'English' coastline is transferred to a modern chart, it appears to coincide with the extreme latitudes of John Cabot's Newfoundland exploration in 1497. But an allowance of two degrees error in latitude determination — over a century later even so accomplished a navigator as Champlain was capable of a forty-minute latitude error — *could* have permitted Cabot a Maine landfall in the next year, a possibility supported by J. A. Williamson and vehemently rejected by Morison. In any case, La Cosa's territorial award to England can scarcely be considered gratuitous whimsy in an age of jealous rivalry for New World claims. Even Harrisse, writing at the close of the nineteenth century, having denied that the evidence supports a second Cabot visit, is compelled to admit that the standards clearly implicate a pre-1500 English expedition to a region that England would not revisit until the middle of Elisabeth's reign.

Appendix 20

The Rock at Grates Cove. Yet Again?

Peter Firstbrook

Note: Peter Firstbrook in his excellent book on John Cabot, The Voyage of the Matthew : John Cabot and the Discovery of North America (1997), makes every attempt to be absolutely complete and comprehensive. To that end, he includes the short section below about the "Cabot Rock" at Grates Cove, and as well he has included a section entitled, "The McCarthy Diary," a reference to Daniel McCarthy, a fictitious character created by Newfoundland writer, Jack Dodd (1902-1978). The "Cabot Rock" story and "The McCarthy Diary" are good items for discussion among amateur historians, but do not fall within the realm of matters that professional historians concern themselves with. The reference to, "a film producer who keeps it at his home in the American mid-west," is really an extension of a folktale gotten out of hand. The legend of the rock is "held dear by the people" of Grates Cove and area, some of whom believe it to be the gospel truth. That in a way is unfortunate, for history cannot and must not be based on legends. Firstbrook says that Pedro Reinel was a Spanish cartographer. Arthur Davies says Reinel was a Portuguese cartographer. This time I am in Arthur Davies' camp.

One of the enduring Newfoundland legends about Cabot is that he died there after his ship was wrecked at Grates Cove, at the eastern end of the island, in 1498. The story has its origins in a map drawn by the Spaniard Pedro Reinel in about 1503. In 1955 a British geographer, Arthur Davies, published a paper in the journal *Nature* claiming that the incident could be reconstructed from details on Reinel's map. Cabot's ship is supposed to have sunk close to Grates Cove and Cabot, his son Sancio and some of the crew swam ashore. A rock on the shore at Grates Cove apparently bore the inscription 'Io. Cabotto' along with others, including 'Sanccius' and 'Sainmalia'; these were interpreted by Davies as a cry for help from the stranded crews — 'Santa Maria save us.'

Davies never actually visited Grates Cove and the rock in question disappeared in the 1960s; some say it was taken by archaeologists, others by a film producer who keeps it at his home in the American mid-west! The Grates Cove legend is a fascinating story that is still held dear by people local to the area, but it is given little credence by historians.

A List of
Sixty
Cabotian Scholars

A List of Sixty Cabotian Scholars

Note: This list of sixty Cabotian scholars includes historians and writers who have researched and written about John Cabot since Richard Biddle (1796-1847) produced the first important secondary work on the Cabot voyages in 1831. For the most part the list includes those who have made some reference to Cabot's voyage of 1498.

I have added the dates or date for each individual, unless I have been unable to ascertain that information. I have also added the country of origin for each individual. In the case of Canada I have added the provinces as well. Newfoundlanders who lived and died before April 1, 1949 had no connection with Canada, because up to that date Newfoundland was one of four nations in North America, and an independent country within the British Empire and Commonwealth.

Here then are the individuals who, in my opinion, may be called Cabotian scholars.

Beazley, Charles Raymond	(1868-1955) - England	
Biddle, Richard	(1796-1847) - U.S.A.	
Biggar, Henry P.	(1872-1938) - Canada — Ontario	
Bourne, Edward G.	(1860-1908) - U.S.A.	
Brown, Rawdon	(1803-1886) - England	
Cram, Fred	(1939-) - Canada — Newfoundland	
Cuthbertson, Brian	(1936-) - Canada — Nova Scotia	
Davies, Arthur	(1906-) - England	
Dawson, Samuel E.	(1833-1916) - Canada — Nova Scotia	
Deane, Charles	(1813-1889) - U.S.A. — Maine	
English, Leo E.F.	(1887-1971) - Canada — Newfoundland	
Fergusson, Charles Bruce	(1911-1978) - Canada — Nova Scotia	
Firstbrook, Peter	( ) - England	
Fraser, Allan M.	(1906-1970) - Scotland — Nfld	
Fuson, Robert H.	( ) - U.S.A. — Florida	
Ganong, William F.	(1864-1941) - Canada — N. Brunswick	
Graham, Gerald S.	(1903-1988) - Canada — Ontario	
Harris, Leslie	(1929-) - Canada — Newfoundland	
Harrisse, Henry	(1829-1910) - France and U.S.A.	
Hoffman, Bernard G.	(1925-) - U.S.A.	
Howley, James P.	(1847-1918) - Newfoundland	
Howley, Michael F.	(1843-1906) - Newfoundland	
Hubbard, Jake T.W.	( ) - U.S.A	
Innis, Harold A.	(1894-1952) - Canada — Ontario	

Jackson, Melvin H.	(1942-) -	U.S.A.
Juricek, John T.	(1938-) -	U.S.A.
Kermode, Lloyd E.	( ) -	U.S.A.
Layng, Theodore E.	(1914-1988) -	Canada — Ontario
Little, Bryan	(1913-) -	England
McGrath, Patrick	( ) -	England
Morison, Samuel E.	(1887-1976) -	U.S.A.
Morris, Roger	(1949-) -	U.S.A. — Maine
Morton, William L.	(1908-1980) -	Canada — Manitoba
Munn, William A.	(1864-1940) -	Newfoundland
O'Dea, Fabian	(1918-) -	Canada — Newfoundland
Oleson, Tryggvi J.	(1912-1963) -	Canada — Manitoba
Olson, Julius E.	(1858-1944) -	U.S.A.
Parsons, John	(1939-) -	Canada — Newfoundland
Pope, Peter E.	(1946-) -	Canada — Newfoundland
Prowse, Daniel W.	(1834-1914) -	Newfoundland
Prowse, George R.F.	(1860-1946) -	Newfoundland — Canada
Quinn, David B.	(1909-) -	England
Rothney, Gordon O.	(1912-1998) -	Canada — Quebec
Rowe, Frederick W.	(1912-1994) -	Canada — Newfoundland
Sacks, David H.	( ) -	U.S.A.
Samuelson, Karl	(1956-) -	Canada — Newfoundland
Saunders, Robert	(1891-1966) -	Newfoundland — U.S.A.
Seaver, Kirsten A.	(1934-) -	U.S.A.
Shaw, Barbara E.	(1937-) -	Canada — Ontario
Simons, Eric N.	(1896-) -	England
Skelton, Raleigh A.	(1906-1970) -	England
Sullivan, Arthur M.	(1932-) -	Canada — Newfoundland
True, David O.	(1885-1967) -	U.S.A.
Vigneras, Louis-Andre	(1903-) -	U.S.A.
Whiteley, George C.	(1908-1990) -	U.S.A. — Newfoundland
Williams, Alan F.	(1930-) -	England
Williamson, James A.	(1886-1964) -	England
Wilson, Ian	(1941-) -	England
Winship, George P.	(1871-1952) -	U.S.A.
Winsor, Justin	(1831-1897) -	U.S.A.

Brief Biographies of
Twenty
Cabotian Scholars

Brief Biographies of Twenty Cabotian Scholars

Note: This list for the most part includes individuals who have researched and written about John Cabot's voyage of 1498. Biographies of Newfoundland Cabotian scholars can be found in the Encyclopedia of Newfoundland and Labrador. Some others can be found in The Canadian Encyclopedia (1988) and other reference works. There are excellent essays on Henry Harrisse and Samuel E. Morison. Please see section C of my John Cabot Bibliography.

Henry P. Biggar (1872-1938) was born in Carrying Place, Ontario. He was a prominent Canadian historian and archivist. For many years he was associated with the National Archives of Canada. He was an authority on early European voyages to Canada. In 1903 he published *The Voyages of the Cabot's and of the Corte-Reals to North America and Greenland, 1497-1503* and in 1911 he published *The Precursors of Jacques Cartier 1497-1534, A Collection of Documents Relating to the Early History of the Dominion of Canada.* This book contains a number of important documents related to John Cabot's voyages in 1497 and 1498.

Fred Cram (b. 1939) is a native of Old Perlican, Trinity Bay. A graduate of Memorial University of Newfoundland, Cram taught school in Newfoundland for over thirty years. A civic-minded individual and a former Mayor of Old Perlican from 1972-1981, Cram may be considered as the chief proponent of the Cabot Rock story in Newfoundland, and therefore a supporter of Arthur Davies' theory that John Cabot was lost in Baccalieu Tickle in December 1498. Cram's recent book on Old Perlican entitled, *As We Were: An Historical Perspective of Old Perlican* (1996), makes reference to John Cabot but does not detail the Cabot Rock story. This writer's ideas about Cabot have been published in newspapers and periodicals.

Brian Cuthbertson (b. 1936) is a native of Nova Scotia and a former Head of Heritage for Nova Scotia. He has written several books pertaining to the history of Nova Scotia including a history of the town of Wolfville, the site of Acadia University. Cuthbertson's book, *John Cabot and the Voyage of the Matthew* (1997) is well-written and well-illustrated. The writer deals briefly with Cabot's voyage of 1498 indicating that Cabot may have encountered the Spanish in the Caribbean Sea. He does not mention Arthur Davies and the Cabot Rock theory. Cuthbertson's book has most of the essential information about Cabot. It is a scholarly work by a competent scholar and is highly recommended.

Arthur Davies (b. 1906) was Reardon-Smith Professor of Geography, University of Exeter, Devon, England from 1948-1971. Davies, who served in the British Army during World War II, is recognized as one of the top scholars in his field in this century. Davies is/was particularly interested in the English discovery and exploration of North America. In 1955 he published an essay in *Nature* magazine entitled, "The Last Voyage of John Cabot and the Rock at Grates Cove." This historical reconstruction is directly related to the legend of the Cabot Rock, a large stone with certain engravings that was once located in the settlement of Grates Cove, Newfoundland. Some people consider Davies' theory as the gospel truth. Davies' theory, however, is a weak argument for associating John Cabot with Grates Cove. Professional historians have never taken Davies' theory seriously.

Leo E.F. English (1887-1971) was born in Jobs Cove, Conception Bay, Newfoundland not far from Grates Cove. A dedicated antiquarian and local historian, English was not always consistent in his opinions about the Cabot Rock at Grates Cove. He taught school in Newfoundland before his appointment as curator of the Newfoundland Museum in 1946, a position he held until the 1960's. English authored or co-authored several booklets related to Newfoundland history, and was a frequent contributor to the local daily newspapers and *The Newfoundland Quarterly*. Initially a strong supporter of the Cabot Rock legend, he tried to evade the subject in his later years. English was a Newfoundland nationalist who believed that logic if nothing else dictated that Cabot in 1497 landed at Cape Bonavista. Logic, however, is no substitute for historical truth.

Peter Firstbrook (b. 1954), an English writer and television producer, has produced a television program about the voyage of the *Matthew*. Firstbrook is an experienced sailor with a degree in oceanography. He was one of the organizers behind the International Festival of the Sea in Bristol in 1996. A frequent writer for newspapers and magazines, Firstbrook is a resident of Bristol. His book on Cabot published in 1997 and entitled, *The Voyage of the Matthew: John Cabot and the Discovery of North America* has been called "the best book to come out during the 500th anniversary year of Cabot's voyage" (Olaf U. Janzen). This is debatable; nevertheless, Firstbrook's work is well-researched and well-illustrated and deals with Cabot's voyages of 1497 and 1498. His section on the 1498 voyage supports the theory that Cabot could have been eliminated by the Spanish somewhere on the coast of South America. This author makes reference to Arthur Davies' theory concerning the Cabot Rock at Grates Cove and as well The McCarthy Diary folktale found in the writings of Newfoundland writer, Jack Dodd (1902-1978).

Allan M. Fraser (1903-1969), historian and archivist, was born in Inverness, Scotland and came to Newfoundland in the late 1920's. For many years he was Head of the History Department at Memorial University College and later at Memorial University of Newfoundland. He served a term as a Member of Parliament for Newfoundland in the 1950's, and in 1958 he became the Chief archivist of the Provincial Archives of Newfoundland. Fraser had little sympathy for those who believed that the Cabot Rock at Grates Cove had a direct connection with John Cabot; consequently, he was not impressed by Arthur Davies' article about John Cabot in *Nature* magazine in November 1955. He wrote several letters to scholars stating his position on this subject. Fraser, who probably never saw the rock, claimed that the engravings could have been put there by the actions of nature. Fraser was wrong. The engravings on the so-called Cabot Rock were put there by someone with hammer and chisel.

Samuel E. Morison (1887-1976), a world famous American historian, was born in Boston and for over forty years was associated with Harvard University. A Rear-Admiral in the United States Navy, Morison had a particular interest in the European discovery of America. His biography of Columbus published in 1942 is entitled, *Admiral of the Ocean Sea*. His two books, *The European Discovery of America: The Northern Voyages, A.D. 500-1600*, and *The European Discovery of America: The Southern Voyages, A.D. 1492-1616* were published in 1971 and 1974. Morison rejected the views of some historians that Cabot landed in New England in 1497, and in 1498 might have gone as far south as the Caribbean Sea. He was adamant in his rejection of the theories of James A. Williamson and Raleigh A. Skelton. Olaf U. Janzen, a professor of history at Sir Wilfred Grenfell College in Corner Brook, Newfoundland, says that Morison in his books, "combines the inquiring mind of the scholar with the nautical skills of an experienced sailor to present convincing analyses of all of the major voyages of exploration which brought Europeans to Newfoundland." (A Cabot Miscellany (1997), page 28). *The Great Explorers*, a good summary of his earlier books on the European discovery of America, was published posthumously in 1978.

William A. Munn (1864-1940) was a prominent Harbour Grace

businessman and well-known writer and historian. It was Munn who first concluded (1914) that the Vikings landed at L'Anse aux Meadows. Munn was a true Cabotian scholar who could not conceive that Cabot landed anywhere other than in Newfoundland. An article about Cabot's landfall in *The Newfoundland Quarterly* in July 1936 indicates that Munn had done considerable research on John Cabot. Munn blamed most of the confusion about Cabot's landfall on Sebastian Cabot and the Cabot map of 1544. He was a true disciple of Prowse and Howley. In 1905 Munn determined once and for all the letters and numbers on the so-called Cabot Rock at Grates Cove. His findings were published in *The Newfoundland Quarterly*. Munn proved that the engravings or markings on the rock had nothing to do with Cabot's voyages in either 1497 or 1498.

Peter E. Pope (b. 1946) is a Newfoundland scholar on the faculty of Memorial University of Newfoundland in St. John's. He is a member of the Archaeology Unit, Department of Anthropology. Pope's book, *The Many Landfalls of John Cabot* was published in Toronto in late 1997. The author is mainly concerned with the debate about Cabot's landfall in 1497, and the way the various landfall theories have been influenced by nationalism or regionalism rather than historical truth. Pope's book which is well-written is an outstanding piece of research by a competent scholar. This book is a must for anybody studying Cabot.

George R. F. Prowse (1860-1946) was born at St. John's, son of historian Daniel W. Prowse. He is referred to in the *Encyclopedia of Newfoundland and Labrador* as a historical cartographer. From the 1890's until his death Prowse was pre-occupied with John Cabot's landfall in 1497 which he argued was in Newfoundland. During the 1920's and 30's he delivered lectures on Cabot in several Canadian cities. He had some of his lectures printed and distributed to various libraries. Prowse's lectures and his letters to Canadian historians, particularly William Ganong, give the impression that he was trying to be objective about Cabot's connection with Newfoundland; whereas in actual fact, he could not see beyond the weak arguments for a landfall at Cape Bonavista put forth by his father in the 1890's. Prowse, who spent most of his life in Winnipeg, died there in 1946. In at least two of his lectures Prowse attempted to authenticate the Cabot Rock legend at Grates Cove.

David B. Quinn (b. 1909) was born in Belfast and graduated from Queen's University, Belfast and the University of London. From 1957-76 Quinn was a professor of history at the University of Liverpool. Quinn has written extensively on the European discovery and exploration of North America. There have been several books and many articles in academic journals. Quinn's book, *North America from Earliest Discovery to First Settlements: The Norse Voyages to 1612* is an excellent source for anyone studying the Cabot voyages and another work edited by Quinn, *New American World: A Documentary History of North America to 1612* is a thorough collection of documents pertaining to the period covered. Most of the Cabot documents are included in this book.

Gordon O. Rothney (1912-1998) was a prominent Canadian historian. He was head of the Department of History at Memorial University of Newfoundland when Arthur Davies wrote his famous or infamous article about the Cabot Rock at Grates Cove. Rothney and his colleague, Allan M. Fraser, dismissed Davies' attempt to reconstruct Cabot's 1498 voyage, and the theory that Cabot was shipwrecked in Baccalieu Tickle. Both Rothney and Fraser wrote letters to scholars in England and the United States stating their position. At one point Rothney got himself into difficulties with the Newfoundland establishment by claiming that Mason's map was a weak argument for Cabot's landfall being at Cape Bonavista. Being a non-Newfoundlander, Rothney was one of the first scholars in modern times to take an objective approach as regards Cabot's landfall. This historian believed that Cabot in 1497 landed somewhere on the coast of Nova Scotia.

Frederick W. Rowe (1912-1994) was a well-known Newfoundland educator, politician and historian. His *History of Newfoundland and Labrador* (1980) is a modern standard study which is complete right up to modern times. Rowe's treatment of Cabot is clear and straightforward and is an excellent summary of the basic research on the Cabot voyages. Everyone interested in Cabot would do well to read the section on John Cabot in this book.

Barbara E. Shaw (b. 1937) was born at Galt, Ontario and graduated from the State University of New York. She completed teacher training in Ontario and is certified in both Ontario and Newfoundland. An artist, musician, poet and writer, she has been a resident of Grates Cove since 1993, and a former chairperson of The Cabot Rock Heritage Corporation at Grates Cove. Shaw, a firm believer in the authenticity of the so-called Cabot Rock, wrote the *Town Crier* speech first used at the Cabot Rock celebrations in 1995. This speech entitled "The True History of Giovani Caboto, Grates Cove, and the Cabot Rock" has been privately printed and distributed. The speech is based on the article by Arthur Davies published in *Nature* magazine in 1955. Shaw believes that the engravings on the rock once located at Grate Cove are authentic. In her own words: "My actual thinking on the Cabot Rock is that nothing can be proven; but, I seriously wonder where its story could have sprung from, if not from some reality." (From a letter to the author dated March 23, 1998).

Raleigh A. Skelton (Peter to his close friends and colleagues) (1906-1970) was born in London. He served in the British Army during the Second World War. A graduate of Cambridge University and a man with an international reputation as a cartographer, he was for many years associated with the Map Room of the British Museum. He was secretary of the Hakluyt Society from 1946 to 1966. Skelton who did the cartographical work on Cabot's voyages, an essay on which was printed in James A. Williamson's book in 1962, was also a major contributor to the Yale University Press volume, *The Vinland Map* (1965). He also did research in cartography associated with the voyages of Magellan and Captain James Cook. Skelton, whose last honorary degree was conferred by Memorial University of Newfoundland in May 1968, died on December 7, 1970 from injuries received in a road accident.

Arthur M. Sullivan (b. 1932) was born in Trinity, Trinity Bay, Newfoundland. A psychologist by training, Sullivan, a Newfoundland Rhodes Scholar (1957) was on the faculty of Memorial University of Newfoundland for over thirty years and served in several senior academic positions including Principal of the Sir Wilfred Grenfell College in Corner Brook. In 1977 he headed a Commission of Inquiry into Newfoundland's transportation system which in 1978 produced a two-volume report. In recent years Sullivan has delved into historical subjects, and has become active as a tour guide and tour organizer. He is head of a company known as Discovery Tourist Services, Incorporated.

Alan F. Williams (b. 1930) is an English geographer and historian who was during the 1960's a member of the Geography Department at Memorial University of Newfoundland. In 1997 he retired from the Department of American and Canadian Studies at the University of Birmingham, a position he had held for over thirty years. His book on John Cabot entitled *John Cabot and Newfoundland* (1996) written under the auspices of the Newfoundland Historical Society summarizes the essentials of the Cabot voyages but takes no position as to a landfall in 1497, nor does Williams get much into Cabot's second voyage. Effectively, this book is a superficial study of Cabot. The book is valuable as a starter in Cabotian studies, but serious students need to move on to more detailed studies; for example, Williamson, Pope, Wilson or Firstbrook.

James A. Williamson (1886-1964) was born in Chichester England and died there 78 years later. He was a well-known naval historian, and one time vice-president of both the British Historical Association and the Hakluyt Society. Williamson was a school teacher from 1910 to 1937 with the exception of four years he served in the British Army during World War I. Associated with Cambridge University after 1937, Williamson's two books on John Cabot are recognized standard works particularly his 1962 book entitled, *The Cabot Voyages and Bristol Discovery under Henry VII*. Anybody studying Cabot must start with this book. This work contains a large number of documents related to John Cabot's voyages of 1497 and 1498.

Ian Wilson (b. 1941) was born in London and graduated from Magdalen College, Oxford, in 1963. A religious scholar as well as a historian, Wilson's first book, *The Turin Shroud* (1978) was an international best-seller. Other religious books include *Jesus: The Evidence* and *The After Death Experience*. *The Columbus Myth* (1991) is a well-known book, in which Wilson claims that John Cabot was really the true discoverer of North America. In this book Wilson supports the theory of James A. Williamson that Cabot went south and encountered the Spanish in the Caribbean Sea. Wilson's most recent book, *John Cabot and the Matthew* (1996) is a kind of summary of his 1991 book in which he claims that John Cabot landed in New England in 1497 possibly as far south as Cape Cod.

The Cabot Monument at Grates Cove, Newfoundland. The story of the "Cabot Rock" at Grates Cove has fascinated amateur historians for generations.

— Courtesy *The Compass*

A John Cabot Bibliography

A. Books and Booklets
B. Articles, Essays, Letters, Theses and Lectures
C. Encyclopedias and Dictionaries
D. Other Bibliographical Material

A: Books and Booklets

Almagia, Roberto, *Commemorazione di Sebastiano Caboto nel IV centenario della morte*, Venice, 1958.

_____ *Gli italiani, prime esploratori della america*, Rome, 1937.

Alusio, Francesco, *Giovanni Cabaoto (John Cabot): A Passion for Discovery*, Toronto, 1997.

Andrews, K.R., N.P. Canny and P.E. Hair (eds.), *The Westward Enterprise: English Activities in Ireland, the Atlantic, and America 1480-1650*, Liverpool, England, 1978.

Anspach, Lewis A., *A History of the Island of Newfoundland*, London, 1819, (Second Edition, London 1827),

Babcock, William H., *Legendary Islands of the Atlantic*, New York, 1922.

Bakeless, John Edwin, *The Eyes of Discovery: The Pageant of North America as seen by the First Explorers*, Philadelphia, 1950.

Ballesteros y Beterra, Antonio, *La marina cantabra y Juan de la Cosa*, Santander, Spain, 1954.

Barrett, William, *The History and Antiquities of the City of Bristol*, Bristol, 1789.

Beazley, C. Raymond, *John and Sebastian Cabot, the Discovery of North America*, London, 1898.

Becher, A.B., *Navigation of the Atlantic Ocean* (5th Edition), London, 1892.

Berger, Josef and Lawrence C. Wroth, *Discoverers of the New World*, New York, 1960.

Biddle, Richard, *A Memoir of Sebastian Cabot, with a Review of the History of Maritime Discovery*, London and Philadelphia, 1831 (Second Edition, 1832).

Biggar, Henry P., *The Precursors of Jacques Cartier 1497-1534, A Collection of Documents Relating to the Early History of the Dominion of Canada*, Ottawa, 1911.

_____ *The Voyages of the Cabots and of the Corte-Reals to North America and Greenland: 1497-1503*, Toronto and Montreal, 1903.

Boorstein, Daniel J., *The Discoverers*, New York, 1983.

Bourne, Edward G. (ed.), "The Voyages of Columbus and of John Cabot," in *The Northmen: Columbus and Cabot*, New York,

1906 (Reprint, 1934).

Bourne, Henry R.F., *English Seamen Under the Tudors*, London, 1868.

Brebner, John B., *The Explorers of North America*, 1492-1806, London, 1933.

Briffett, Frances B., *The Story of Newfoundland and Labrador*, Toronto, 1949 (Reprint, 1964).

Briggs, Lloyd V., *History and Genealogy of the Cabot Family, 1475-1925*, Boston, 1927.

Brown, Richard, *A History of the Island of Cape Breton with Some Account of the Discovery and Settlement of Canada, Nova Scotia and Newfoundland*, London, 1869.

Bulgin, Iona and Bert Riggs (eds.), *A Cabot Miscellany,* The Newfoundland Historical Society, St. John's, 1997.

Bunbury, Edward H., *A History of Ancient Geography*, New York, 1959.

Burpee, Lawrence J., *The Discovery of Canada*, Toronto, 1944.

Campbell, John, *Lives of the Admirals and other Eminent British Seamen*, London, 1750.

Candow, James E., *Daniel Woodley Prowse and the Origin of Cabot Tower* (Parks Canada, No. 155), Ottawa, 1981.

Cardi, Luigi, *Gaeta partria di Giovanni Caboto*, Roma, 1956.

Carus, Edward M., *The Merchant Adventurers of Bristol in the Fifteenth Century*, Bristol, 1962.

Carus-Wilson, Eleanora Mary, *The Merchant Adventurers of Bristol in the Fifteenth Century*, London, 1962.

_____ *Medieval Merchant Venturers*, London, 1954

Chadwick, St. John, *Newfoundland, Island into Province*, Cambridge, England, 1967.

Channing, Edward (ed.), *Documents Describing the Voyages of John Cabot in 1497*, New York, 1893.

Cochrane, James A., and A.W. Parsons, *The Story of Newfoundland*, Toronto, 1949.

Costello, Kieran, *John Cabot's Matthew, 1497-1997: A Voyage from the Past into the Future*, Bristol, 1995.

Cram, E. Fred, *As We Were: An Historical Perspective of Old Perlican, 1568-1968,* St. John's, 1996.

Crone, Gerald R., *Maps and Their Makers: An Introduction to the History of Cartography*, London, 1953 (Revised Edition, 1968 - Reprint 1978).

_____ *Maps and the New World: Cartographical Development 1475-1625*, London, 1962.

_____ *The Discovery of America: Atlantic Voyages of the Portuguese*, New York, 1969.

Cuff, Robert H., *New-Found-Land at the Very Centre of the European Discovery and Exploration of North America*, St. John's, 1997.

Cumming, W.P., R.A. Skelton and D.B. Quinn, *The Discovery of North America*, Toronto and Montreal, 1972.

Cuthbertson, Brian, *John Cabot and the Voyage of the Matthew*, Halifax, 1997.

Daniell, David S., *Explorers and Exploration*, London, 1962.

Dawson, Samuel E., *The Voyages of the Cabots in 1497 and 1498, with an attempt to determine their landfall and to identify their island of St. John*, Ottawa and Montreal, 1894.

_____ *The Voyages of the Cabots: Latest Phases of the Controversy*, Ottawa, Toronto, London, 1897.

Dionne, Narcisse E., *John and Sebastian Cabot*, Quebec City, 1898.

Dodd, Jack, *John Cabot's Voyage to Newfoundland*, St. John's, 1974.

Eden, Richard, *A Treatyse of the Newe India, with other new found landes and Islandes, as well eastwards as westwards*, London, 1553.

_____ *Decades of the New World*, London, 1555.

Elliott, J. H., *The Old World and the New, 1492-1650,* Cambridge, England, 1972.

English, Leo, E.F., *Historic Newfoundland*, St. John's, 1955.

Fardy, Bernard D., *John Cabot: The Discovery of Newfoundland*, St. John's, 1994.

Fernandez-Armesto, Felipe, *The Times Atlas of World Exploration,* London, 1991.

Fernandez de Navarrete, Martin, *Colleccion de los viajes y descubrimientos que hicieron por mar los Espanoles,* Volume III, Madrid, Spain, 1829.

Firstbrooke, Peter, *The Voyage of the Matthew: John Cabot and the Discovery of North America,* Toronto, 1997.

Fiske, John, *The Discovery of America*, Volume 2, Boston, 1892.

Ganong, William F., *Crucial Maps in the Early Cartography and Place-Nomenclature of the Atlantic Coast of Canada*, Toronto, 1929-1937 (New Edition with an Introduction by Theodore E. Layng, 1964).

Gibbons, Henry K., *The Myth and the Mystery of John Cabot,* Port aux Basques, Newfoundland, 1997.

Goodnough, David, *John Cabot and Son*, Mahwah, New Jersey, 1979.

Gosling, William G., *Labrador — Its Discovery, Exploration and Development*, London, 1910.

Hakes, Harry, *John and Sebastian Cabot, a Four Hundredth Anniversary Memorial of the Discovery of America*, Wilkes-Barre, Pennsylvania, 1897.

Hakluyt, Richard, *Divers Voyages Touching the Discovery of America*, London, 1582.

_____ *The Principel Navigations, Voyages, Traffiques and Discoveries of the English Nation* (Reprint), London, 1907-1910.

Harris, Leslie, *Newfoundland and Labrador — A Brief History*, Toronto, 1968.

Harris, R. Cole, *Historical Atlas of Canada, Volume 1, From the Beginning to 1800*, Toronto, 1987.

Harrisse, Henry, *When did John Cabot Discover North America?*, London, 1897.

_____ *Jean and Sebastian Cabot*, Paris, 1882 (Reprint, Amsterdam, 1968).

_____ *Decouverte et evolution cartographique de Terre-Neuve*, Paris, 1900.

_____ *John Cabot, the Discoverer of North America, and Sebastian his son*, London, 1896 (Reprint, New York, 1968).

_____ *The Diplomatic History of America*, London, 1897.

_____ *The Discovery of North America*, London and Paris, 1892.

_____ *La Decouverte et evolution cartographique de Terre-Neuve et des pays circomvoisins, 1497-1769*, Paris, 1890 (Reprint, Paris and London, 1900).

Hart, Albert B. (ed.), *American History Told by Contemporaries, Volume I (Era of Colonization, 1492-1689)*, New York, 1897.

Harvey, Moses, *Newfoundland in 1897*, London, 1897.

_____ *Textbook of Newfoundland History*, Boston, 1895.

_____ *A Short History of Newfoundland*, London, 1890.

_____ *Documents Describing the Voyage of John Cabot*, New York, 1893.

Hatton, Joseph & Moses Harvey, *Newfoundland, the Oldest British Colony*, London, 1883.

Hill, Kay, *And Tomorrow the Stars: The Story of John Cabot*, New York, 1968.

Hobley, L.F., *Early Explorers to A.D. 1500*, London, 1954.

Hoffman, Bernard G., *Cabot to Cartier: Sources for an Historical Ethnography of Northeastern North America, 1497-1550*, Toronto, 1961.

Horsford, Eben N., *John Cabot's Landfall in 1497 and the site of Norumbega*, Cambridge, Mass., 1886.

Howard, Richard, *Bristol and the Cabots*, Jackdaw, Series No. 69 (Clarke-Irwin and Company), Toronto, 1967.

Howley, James P., *The Beothucks or Red Indians: The Aboriginal Inhabitants of Newfoundland*, Cambridge, England, 1915.

Hughes, Walter W., *A short account of the Cabots and the first discovery of the Continent of America*, Bristol, 1897.

Jacobson, Timothy C., *Discovering America: Journeys in Search of The New World*, Toronto, 1991.

Johnson, Adrian Miles, *America Explored: A Cartographical History of the Exploration of North America*, New York, 1974.

Jones, John W. (ed.) *Divers Voyages Touching the Discovery of American and the Adjacent Islands* (Edited works of Richard Hakluyt, 1582), London, 1850.

Kidder, Frederic, *The Discovery of North America by John Cabot: A First Chapter in the History of North America*, Boston, 1878.

Kohl, Johann G., *A History of the Discovery of the East Coast of North America particularly of the coast of Maine*, Boston, 1869.

_____ *Documentary History of the State of Maine* (edited by William Willis), Volume 1 containing a *History of the Discovery of Maine*, Portland, Me., 1869.

Kurtz, Henry Ira, *John and Sebastian Cabot*, New York, 1973.

Larned, Joseph N. (ed.), *History for Ready Reference* — "Kohl on Cabot," Springfield, Mass., 1894.

Larsen, Egon, *John and Sebastian Cabot*, Stuttgart, Germany, 1985.

Larsen, Sofus, *The Discovery of America Twenty Years Before Columbus*, London and Copenhagen, 1925.

Latimer, John, *History of the Society of Merchant Venturers of the City of Bristol*, 1903.

Layng, Theodore E., *Sixteenth Century Maps Relating to Canada*, Toronto, 1958.

Layng, Theodore E., *Sixteenth Century Maps Relating to Canada*, Ottawa, 1966.

Leacock, Stephen, *The Dawn of Canadian History,* Toronto, 1914. Reprint 1964.

Little, Bryan, D.G., *John Cabot: The Reality*, Bristol, 1983.

Macdonald, Peter, *Cabot and the Naming of America: a Revelation,* Bristol, 1997.

Major, Richard H., *The True Date of the English Discovery of America*, London, 1871.

Marcus, G.J., *The Conquest of the North Atlantic*, New York, 1981.

Markham, Clements R., *The Journal of Christopher Columbus and Documents relating to the Voyages of John Cabot*, London, 1893.

Martyr, Peter, *De rebus Oceanicis et Orbe Nova Tres*, Second Edition, Seville, Spain, 1533.

_____ *Decades of the New World or West India*, Alcalia, Spain, 1516.

_____ *Decades of the Oceans*, Alcalia, Spain, 1516.

Mason, John, *A Briefe Discourse of the New-found-land . . .* , London, 1620.

Mavor, William F., *Historical Account of the Most Celebrated Voyages*, London, 1801.

Maxwell, William F., *The Newfoundland and Labrador Pilot*, (3rd Edition), London, 1897.

McBride, Elizabeth R., *The Story of the Cabots*, Dansville, New York, 1910.

Morison, Samuel E., *The European Discovery of America: The Northern Voyages 500-1600 A.D.*, New York, 1971.

_____ *The European Discovery of America: The Southern Voyages 1492-1616,* New York, 1974.

_____ *The Great Explorers*, New York, 1978.

Morris, Roger, *Atlantic Seafaring: The Centuries of Exploration and Trade in the North Atlantic,* Camden, Maine, 1992.

Newton, Arthur P., *The Great Age of Discovery,* London, 1932.

Nicholls, J.F., *The Remarkable Life, Adventures and Discoveries of Sebastian Cabot,* London, 1869.

Norman Charles, *Discoveries of America,* New York, 1968.

Nuffield, Edward W., *The Discovery of Canada,* Vancouver, 1996.

Nunn, George E., *The mappemonde of Juan de La Cosa: a critical investigation of its data,* Jenkintown, Pennsylvania, 1934.

_____ *The La Cosa Map and the Cabot Voyages,* New York, 1946.

Ober, Frederick A., *John and Sebastian Cabot,* New York, 1908.

Oleson, Tryggvi J., *Early Voyages and Northern Approaches, 985-1600,* Toronto, 1963.

Olson, Julius E., (ed.) *The Voyages of the Northmen,* in *The Northmen: Columbus and Cabot,* New York, 1906 (Reprint, 1934).

_____ "The Voyages of the Northmen," in J.F. Jameson (ed.), *Original Narratives of Early American History,* Volume 1, New York, 1906.

Outwaite, Leonard, *The Atlantic, History of an Ocean,* Toronto, 1958.

Packard, A.S., *The Labrador Coast,* New York, 1891.

Parsons, John, *Labrador — Land of the North,* New York, 1970.

_____ *Away Beyond the Virgin Rocks: A Tribute to John Cabot,* St. John's, 1997.

Payne, Edward T., *History of the New World called America,* Oxford, 1892.

Pedley, Charles, *The History of Newfoundland from the Earliest times to the year 1860,* London, 1863.

Penrose, Boies, *Travel and Discovery in the Renaissance 1420-1620,* Cambridge, Mass., 1952.

Phillips, William D., and Carla R. Phillips, *The World of Christopher Columbus,* Cambridge, England, 1992.

Pollard, A.F., *The Reign of Henry VII,* London, 1914.

Pope, Peter E., *The Many Landfalls of John Cabot,* Toronto, 1997.

Prowse, Daniel W., *A History of Newfoundland from the English, Colonial and Foreign Records,* London, 1895.

Prowse, George R.F., *Cabot's Surveys*, Winnipeg, 1931.

_____ *The Cabot Landfall*, Winnipeg, 1938 (Reprint, 1972).

_____ *John Cabot's Baccalaos: John Cabot named the New Isle Baccalaos*, Winnipeg, 1940.

_____ *Cabot's Bona Vista Landfall*, Winnipeg, 1946.

_____ *Exploration of the Gulf of St. Lawrence 1499-1525*, Winnipeg, 1929.

Quinn, David B., *Sebastian Cabot and Bristol Exploration*, 1968 (Revised Edition, 1993).

_____ *England and the Discovery of America 1481-1620*, London, 1974.

_____ *North America from Earliest Discovery to First Settlements*, New York, 1975 (Second Edition, 1977).

_____ *North American Discovery*, London, 1971.

_____ *New American World: A Documentary History of North America to 1612* (5 volumes), New York, 1979.

_____ *The New Found Land: The English Contribution to the Discovery of North America,* Providence, Rhode Island, 1965.

Rasky, Frank, *The Polar Voyagers*, Toronto, 1976.

Rein, Adolph, *Der Kampf Westeruopas um Nord Amerika in 15 und 16 Jahrundert*, Stuttgart, Germany, 1925.

Robinson, Conway, *An Account of Discoveries in the West until 1519*, Richmond, Virginia, 1848.

Rowe, Frederick W., *A History of Newfoundland and Labrador*, Toronto, 1980.

Sacks, Davis H., *The Widening Gate: Bristol and the Atlantic Economy*, 1450-1700, Los Angeles, 1993.

Samuelson, Karl, *Fourteen men who figured prominently in the story of Newfoundland and Labrador*, St. John's, 1984.

Sauer, Carl O., *The Early Spanish Main*, Berkeley and Los Angeles, 1966.

Scadding, Henry, *Seneca's Prophecy and its Fulfilment*, Toronto, 1897.

Schwartz, Seymour I. and Ralph E. Ehrenberg, *The Mapping of America*, New York, 1980.

Seary, E.R., *Family Names of the Island of Newfoundland*, St. John's, 1977.

Seaver, Kirsten A., *The Frozen Echo: Greenland and the Exploration of North America, ca. A.D. 1000-1500,* Stanford, California, 1996.

Severin, Tim, *The Brendan Voyage,* London, 1978.

Sharp, John J., *Discovery in the North Atlantic,* Halifax, Nova Scotia, 1991.

Sherborne, J.W., *The Port of Bristol in the Middle Ages,* 1965.

Simons, Eric N., *Into Unknown Waters: John and Sebastian Cabot,* London, 1964.

Skelton, Raleigh A., *Explorers' Maps,* London, 1958.

_____ *The European Image and Mapping of America, A.D.1000-1600,* Minneapolis, 1964.

Skelton, Raleigh A., Thomas E. Marston and George D. Painer, *The Vinland Map and the Tartar Relation,* New Haven and London, 1965.

Smith, Ed., *Fish 'n' Chips — A Brief Twisted History of Newfoundland ... Sort of,* Springdale, Newfoundland, 1997.

Story, George M. (ed.), *Early European Settlement and Exploration in Atlantic Canada: Selected papers,* Memorial University, St. John's, 1982.

Strang, Herbert (ed.), *The Great Explorers,* London, 1934.

Syme, Ronald, *John Cabot and his son Sebastian,* New York, 1972.

Tarducci, Fransceso, *Di Giovanni e Sebastiano Caboto: memorie raccolte e documentate, Venezia,* 1892 (For English translation see Brownson, H.F., Detroit, 1893).

Taylor, Eva, G.R., *Tudor Geography 1485-1583,* London, 1930.

Thacher, John Boyd, *The Continent of America: Its Discovery and its Baptism,* New York, 1896.

Thomas, A.H., and I.D. Thornley (eds.), *The Great Chronicles of London,* London, England, 1939.

Tocque, Philip, *Wandering Thoughts or Solitary Hours,* Dublin and London, 1846.

_____ *Newfoundland, as it was and as it is in 1877,* Toronto, 1878.

Thomson, Don W., *Men and Meridians,* Ottawa, 1966.

Thompson, Jo-Ann, *John Cabot — Commemorative Postage Stamps,* New York, 1932.

Tytler, Patrick F., *Historical view on the progress of discovery on the more northern Coast of America*, London, 1833.

Vaughan, William, *The Golden Fleece*, London, 1626.

Walsh, John C., *Being a strange account of the voyage by John Cabot*, New York, 1962.

Weare, George E., *Cabot's Discovery of North America*, London, 1897.

Weise, Arthur James, *The Discovery of America to the year 1525*, London and New York, 1884.

West, Delno C., *Braving the North Atlantic*, New York, 1996.

Whiffen, Bruce L., *Prime Berth*, St. John's, 1993.

Whitbourne, Richard, *A Discourse and Discovery of the New-foundland*, London, 1620.

Williams, Alan F., *John Cabot and Newfoundland*, St. John's, 1996.

Williamson, James A., *The Voyages of the Cabots and the English Discovery of North America under Henry VII and Henry VIII*, Cambridge, England, 1929.

_____ *The Ocean in English History (The Ford Lectures)*, Oxford, 1941.

_____ *Maritime Enterprise, 1485-1558*, New York, 1972.

_____ *The Cabot Voyages and Bristol Discoveries Under Henry VII* (from the Hakluyt Society Series, Number 120), Cambridge, England, 1962.

Wilson, Ian, *The Columbus Myth: Did Men of Bristol Reach America before Columbus*, London, 1991.

_____ *John Cabot and the Matthew*, Bristol and St. John's, 1996.

Winship, George P., *Cabot Bibliography*, London and New York, 1900 (Reprint, 1967).

_____ *Some Facts About John and Sebastian Cabot*, Worcester, Mass., 1900.

Winsor, Justin, *The Cabot Controversies and the right of England to North America*, Cambridge, Mass., 1896.

Wissler, Clark, Constance L. Skinner and William Wood, *Adventurers in the Wilderness, Vol. 1 of The Pageant of America Series* (ed.), Ralph H. Gabriel (ed.), New Haven, Toronto and London, 1925.

Wood, Herbert J., *Exploration and Discovery*, London, 1951.

Woodbury, Charles L., *The Relation of the Fisheries to the Discovery and Settlement of North America*, Boston, 1880.

B: Articles, Essays, Letters, Theses and Lectures

Allen, John L., "From Cabot to Cartier: the Early Exploration of Eastern North America, 1497-1543," *Annals of the Association of American Geographers*, Volume 82, Number 3, September, 1992.

Anderson, Hugh A., "A Message to Newfoundland from John Cabot", *Atlantic Guardian*, Volume 6, Number 3, March, 1949.

Ava, Frederick Temple Hamilton-Temple Blackwood (1st Marquis of Dufferin), "John Cabot", *Scribner's Magazine*, Volume 22, July, 1897.

Babcock, William H., "Legendary Islands of the Atlantic", the *American Geographical Society Publication*, Research Series, Number 8, New York, 1922.

Baker, George S., "John Cabot and the Discovery of Newfoundland", *Nautical Magazine*, June 1897, also in *Scientific American, supplement*, New York, October 16, 1897.

Ballesteros-Gaibrois, M., "Juan Caboto en Espana, *Revista de Indios,* Volulme 4, 1943.

Barnard, Murray, "For the sake of argument: John Cabot discovered Canada. So what do we do? Ignore him.", *Maclean's Magazine*, Volume 80, Number 4, April, 1967.

Baxter, James P., "John Cabot and his Discoveries", in the *Proceedings* of the Maine Historical Society, Portland, Me., October, 1897.

Beaudovin, Joseph D., "John Cabot", *L'evis*, 1898.

Beaugrand-Champagne, Artistide, "Introduction aux voyages de Jacques Cartier and Jean Cabot", the *Canadian Historical Association*, Annual Report, 1935.

Benson, Bob, "Hard Feelilngs Over Missing Rock," *The Evening Telegram,* Volume 118, Number 239, December 1, 1996.

Biggar, Henry P., "Voyages of the Cabots and the Corte-Reals to North America and Greenland, 1497-1503," *Review of Historical Publications Relating to Canada*, Volume 10, 1895.

_____ "The Voyages of the Cabots and the Corte-Reals", *Revue Hispanique*, Volume 10, 1903.

_____ "The First Explorers of the North American Coast", in A.P. Newton (ed.) *The Great Age of Discovery*, London, 1932.

_____ "A Cabot source which does not exist", Extrait de la *Revue Hispanique*, Volume 15, New York and Paris, 1906.

Black, James W., "The Old World at the Dawn of Western Discovery", in the *Proceedings* of the Maine Historical Society, Portland, Me., June, 1897.

Bond, C.C.J., "The Mapping of Canada, 1497-1658," *Canadian Geographical Journal,* August, 1965.

Bonnycastle, Richard H., "Historical Memoir of Cabot, the Discoverer of Newfoundland", in *Newfoundland in 1842*, London, 1842.

Bosa, Peter, "Perche e importante conoscere la nostra storia: Giovanni Cabot", *Veltro*, Italy, Volume 29, Numbers 1 and 2, 1985.

Bourinot, John G., "The Cabot celebrations in Nova Scotia", in *The Independent*, New York, June 24, 1897.

Brevoort, James C., "Early Voyages from Europe" (John Cabot's Voyages of 1497), in *The Historical Magazine, Morrisania*, New York, March, 1868.

Brown, Stuart C., "For Other Worlds and Other Seas: the context of claims for pre-Columbian European contact with North America," *Newfoundland Studies,* Volume 9, Fall 1993.

Browning, Thomas B., "Discovery of Newfoundland", *The Evening Telegram*, St. John's, June 4, 1928.

Burpee, Lawrence J., "John Cabot, who sought Cipangu, and found Canada", *Canadian Geographical Journal*, Volume 6, Number 6, June, 1933.

Burrage, Henry S., "The Cartography of the Period," in the *Proceedings* of the Maine Historical Society (Cabot Meeting), Portland, Me., June, 1897.

Campeau, Lucien, "Jean Cabot et al descourverte de l'Amerique du Nord," *Revue d'Histoire de l'Amerique Francaise*, Volume 19, 1965.

155

Candow, James, "Daniel Woodley Prowse and the Origin of the Cabot Tower," *Research Bulletin,* Number 155, Ottawa, 1981.

Carter, Harry, "Cabot Discovery," *Newfoundland Stories and Ballads*, Volume 2, Number 2, Autumn, 1955.

Carus-Wilson, Eleanora Mary, "The Overseas Trade of Bristol in the Later Middle Ages," *Bristol Record Society's publications*, Volume 7, Bristol, 1937.

Cole, D.S., "Adventurers of Bristol," *Canadian Geographical Journal*, Volume 6, Number 2, February, 1933.

Crone, Gerald R., "The Vinland Map Cartographically Considered," *Geographical Journal*, 132, 1966.

Cuthbertson, Brian, "John Cabot and His Historians: Five Hundred Years of Controversy," *Lecture to the Nova Scotia Historical Society,* Halifax, May 15, 1996.

Dawson, Nick, "The Real Legacy of John Cabot," *The Evening Telegram,* Volume 119, Number 113, July 26, 1997.

Dawson, Samuel E., "Voyages of the Cabots," A Paper from the *Transactions* of the Royal Society of Canada in 1896 with Appendices on kindred subjects, Ottawa, 1896.

_____ "The Voyages of the Cabots in 1497 and 1498: A sequel to a paper in the "Transactions" of 1894," in the *Transactions* of the Royal Society of Canada, Volume 2, Series 2, 1896.

_____ "The Discovery of America by John Cabot in 1497," being extracts from the *Proceedings* of the Royal Society of Canada, with appendices on kindred subjects, Ottawa, 1896.

Davies, Arthur, "The "English" coasts on the map of Juan de la Cosa," *Imago Mundi*, Volume 13, 1956.

_____ "The Last Voyage of John Cabot," *Nature*, Volume 176, 1955.

Deane, Charles, "The Voyages of the Cabots" in Justin Winsor (ed.), *Narrative and Critical History of America*, Volume 3, Chapter 1, New York 1884-1889 (Reprint, 1967).

_____ "Remarks of Dr. Deane on Sebastian Cabot's mappemonde," in the *Proceedings* of the American Antiquarian Society, Cambridge, Mass., April, 1867.

Dionne, Narcisse E., "John and Sebastian Cabot," in *Le Couvrier du Livre Canadiana*, Quebec City, May-June, 1898.

Djwa, Sandra, "Early Explorations: New Founde Landys, 1496-1729," *Studies in Canadian Literature,* Volume 4, Summer 1979.

Dunbabin, Thomas, "Venice and the Discovery of Canada," *Atlantic Advocate*, Volume 50, Number 10, June, 1960.

_____ "Seven-born John Cabot," *Dalhousie Review*, Volume 42, Autumn, 1962.

_____ "Cabot's Landfall in North America," *Atlantic Advocate*, Volume 48, Number 10, June, 1958.

Dyer, Bruce, "Was John Cabot really the first trans-Atlantic visitor to Canada?," *Canada and the World*, Volume 43, Number 9, May, 1978.

_____ "To the Edge of the Earth" in *Canada and the World*, Volume 43, Number 9, May, 1978.

Earle, Neil, "Cabot's Voyage Matter of Interpretation: One of the oldest Debates in Canada," *The Compass*, Volume 28, Number 32, Carbonear, Newfoundland, August 13, 1996.

Elliot, Marion, "Cabot's visit to Newfoundland in 1497 was a trading venture," *Maritime Advocate and Busy East*, Volume 43, Number 7, March, 1953.

English, Leo, E.F., "The Land First Seen," *Newfoundland Stories and Ballads*, Volume 8, Number 2, St. John's, 1962.

_____ "Cabot's Course and Landfall," *Newfoundland Stories and Ballads*, Volume 13, Number 2, Spring, 1967.

Fergusson, C. Bruce, "Cabot's Landfall," *Dalhousie Review*, Volume 33, Number 4, Winter, 1953.

_____ "Cabot and Cape Breton Island," *Maritime Advocate and Busy East*, Volume 43, Number 9, May, 1953.

Ferguson, G.D., "The Cabots and the Discovery of Canada," *Queen's Quarterly*, Volume 5, Number 2, October, 1897.

Fleming, Stanford, "Canada and Ocean Highways," in *The Journal of the Royal Colonial Institute*, Montreal, 1892, London, 1896.

Flynn, Mike, "Cabot died here, says man," *The Evening Telegram*, St. John's, February 8, 1995.

Fowler, John, "John Cabot was Here" in *The Compass*, September 17, 1996.

Fuson, Robert H., "The John Cabot Mystique," in Stanley H. Palmer and Dennis Reinhartz (eds.) *Essays on the History of North American Discovery and Exploration* (A Walter Prescott Webb Memorial Lecture), Arlington, Texas, 1988.

Gallo, Rodalfo, "Intorno a Giovanni Caboto," *Rendiconti Accod. Naz dei Lincei,* Volume 3, 1948.

Ganong, William F., "The Cartography of the Gulf of St. Lawrence" in the *Transactions* of the Royal Society of Canada, Ottawa, 1890.

Gard, Peter, "Newfoundlander needs a reason to celebrate," *The Newfoundland Herald*, November 19, 1994.

Gardner, Ralph W., "How John Cabot Didn't Discover Cape Breton," in *The Atlantic Advocate*, June, 1977.

Gilchrist, John H., "Cabotian conjectures: did Cabot reach Maine in 1498?," *American Neptune*, Volume 45, Number 4, 1985.

Granger, Charles R., "John Cabot's soliloquy," *Atlantic Guardian*, Volume 7, Number 6, June, 1950.

Graham, Gerald S., "Newfoundland in British Strategy from Cabot to Napoleon" in R.A. Mackay, *Newfoundland Economic, Diplomatic, and Strategic Studies*, Toronto, 1946.

Guy, Allan R., "The 'Cabot Rock' at Grates Cove," in the *Atlantic Guardian*, Volume 13, Number 4, April, 1956.

Harris, Leslie, "The Cabot Landfall: An Examination of the Evidence," *Lecture* to the Newfoundland Historical Society, October 11, 1967 (Transcript by Arthur Fox).

Harrisse, Henry, "Did Cabot return from his second voyage?," *The American Historical Review*, Volume 3, Number 3, April, 1898.

_____ "The Outcome of the Cabot Quarter-Centenary," *The American Historical Review*, Volume 4, Number 1, October, 1898.

_____ "The Cabots" in the *Transactions* of the Royal Society of Canada, Ottawa, 1898.

Hart, Tracy, "John Cabot: How it all Began," *English 2010 paper*, Memorial University of Newfoundland, 1992.

Harvey, Moses, "Voyages and Discoveries of the Cabots," *Nova Scotia Historical Society, Report and Collections, 1893-1895*, Volume 9, Halifax, 1895.

_____ "The Discoverer of North America and the first colonizer of Newfoundland," *Maritime Monthly*, Volume 4, Number 4, Boston, October, 1874.

Hermannsson, Halldor, "The Northmen in America 982-1500," - *Islandica*, Volume II, Ithaca, New York, 1909.

Hill, Kay, "Three Kinds of Research," in the *Canadian Author and Bookman,* Volume 46, Number 4, Summer, 1971.

Howley, James P., "The Landfall of Cabot," in the *Bulletin-Transactions* (1886-1889) of the Geographical Society of Quebec, Quebec City, 1889.

Howley, Michael F., "Cabot's Landfall," in the *Magazine of American History,* New York, October, 1891.

_____ "Cabot's Voyages" — A *lecture* delivered at St. Patrick's Hall, St. John's, Newfoundland, January 11, 1897.

_____ "Latest Lights on the Cabot Controversy" in the *Transactions* of the Royal Society of Canada, Second Series, Volume 9, 1903.

_____ "Mason's Map and Vaughan's Vagaries," *The Newfoundland Quarterly,* Volume 11, Number 1, July 1911. (A speech delivered in St. John's, April 29, 1911).

Hubbard, Jake T.W., "John Cabot's Landfall: Cape Degrat or Cape Bonavista?," *American Neptune,* Volume 33, Number 3, June, 1973.

Jackman, Leo J., "Cabot," *The Newfoundland Quarterly,* Volume 60, Number 1, St. John's, Spring, 1961.

_____ "Criticism of Cabot puts Cart Before Horse," *The Newfoundland Quarterly,* Volume 58, Number 3, St. John's, September, 1959.

Jackson, Melvin H., "The Labrador Landfall of John Cabot: the 1497 Voyage Reconsidered," *Canadian Historical Review,* Volume 44, Number 2, June, 1963.

Juricek, John T., "John Cabot's first voyage, 1497," *Smithsonian Journal of History,* Volume 2, Number 4, Washington, D.C., 1967/68.

Keller, Allan, "Silent Explorer: John Cabot in North America," *American History Illustrated,* Volume 8, Number 9, 1974.

Kelley, James E., "Non-Mediterranean Influences that Shaped the Atlantic in the Early Portolan Charts," *Imago Mundi,* Volume 31, 1979.

Kermode, Lloyd E., "The Spirit of Adventure: John Cabot, the merchants of Bristol and the Re-discovery of America," *The Beaver,* October-November, 1996.

Kinsella, P.J., "Cabot's Dream," in *Our Country,* May 24, 1907.

Lamontagne, Yves, "History Gives Gaspar Corte Real Credit for

Cabot's Discovery of Newfoundland," *The Newfoundland Quarterly*, Volume 58, Number 3, St. John's, September, 1959.

Layng, Theodore E., "Charting the course [of John Cabot] to Canada," *Congresso Internacional de Historia dos Descobrimentos*, Volume 2, Lisbon, 1961 (Reprint, 1965).

_____ "Introduction, Commentary and Map notes" in W.F. Ganong, *Crucial Maps and the Early Cartography and Place-Nomenclature of the Atlantic Coast of Canada*, Toronto, 1964.

_____ "Highlights in the Mapping of Canada," *Canadian Library*, 1960. Reprinted by the Information Division, Department of External Affairs, Reprint Number 253.

Leckie, Campbell A., "John Cabot's Letters Patent," *The Downhomer*, Volume 9, Number 5, October, 1996.

Macdonald, William, "The Landfall of Cabot and the extent of his Discoveries" in the *Proceedings* of the Maine Historical Society, Portland, Me., Volume 8, 1897.

Mackintosh, Charles H., "Cabot and other Western Explorers" in the *Canadian Magazine*, Toronto, December, 1896.

Major, Alan, "Discoverer of the 'New Founde Land'," *Nautical Magazine*, Volume 254, Number 4, October, 1995.

Major, Richard H., "The true Date of the English Discovery of the American Continent under John and Sebastian Cabot, *Archaelogia*, Volume 13, Boston, 1871.

March, William M., "New Fuel added to the Cabot Controversy?," *Halifax Chronicle Herald*, Volume 10, Number 152, June 26, 1958.

Markham, Clements R., "Fourth Century of the Voyage of John Cabot, 1497" in *The Geographical Journal* of the Royal Geographical Society, London, June, 1897.

_____ "Voyage of John Cabot in 1497," *The Geographical Journal*, Volume 9, Number 6, June, 1897.

McGhee, Robert, "Canada Rediscovered" in *Libre Expression*, Canadian Museum of Civilization, Hull, Quebec, 1991.

_____ "Northern Approaches," *Beaver* Volume 72, June/July, 1992.

McGrath, Patrick, "Bristol and America, 1480-1631" in Andrews, K.R. et al (eds.), *The Westward Enterprise: English Activities in Ireland, the Atlantic, and America, 1480-1650*, Liverpool, 1978.

Mead, Edwin D., "The Voyage of the Cabots," in *Old South Leaflets*, Number 37, Boston, 1895.

Morison, Samuel E., "Cabot, the mysterious sailor who gave England rights to North America," *Smithsonian*, Volume 2, Number 1, April, 1971.

Munn, William A., "John Cabot's Landfall," *The Newfoundland Quarterly*, Volume 36, Number 1, St. John's, July, 1936.

Murphy, Michael P., "Voyage of Discovery: Cabot and Newfoundland 1497," *Atlantic Guardian*, Volume 13, Number 6, June, 1965.

Nordenskiold, A.E., "The La Cosa, Cabot and Michael Lok Maps," in *Periplus*, Volume 43, Stockholm, 1897.

Norman, William, "Stamps of Newfoundland: (Cabot)," *Atlantic Guardian*, Volume 4, Number 1, October, 1947.

_____ "Stamps of Newfoundland: The Cabot Issue of 1897," *Atlantic Guardian*, Volume 4, Number 2, Atlantic Guardian, November, 1947.

O'Brien, Cornelius, "Presidential Address on Cabot's Landfall," in the *Proceedings* of the Royal Society of Canada, Ottawa, 1897.

O'Dea, Fabian, "Cabot's Landfall — Yet Again," *The Newfoundland Quarterly*, Volume 68, Number 3, St. John's, Fall, 1971.

O'Dea, Shane, "Judge Prowse and Bishop Howley: Cabot Tower and the Construction of Nationalism," *Lecture to the Newfoundland Historical Society,* St. John's, April 25, 1996.

Oleson, Tryggvi J. and William L. Morton, "Northern Approaches," in *Dictionary of Canadian Biography*, Volume I, Toronto, 1966.

Parsons, John, "John Cabot," *The Canadian Encyclopedia*, Edmonton, 1988.

_____ "It's a Big Lie — John Cabot did not discover Newfoundland in 1497," *The Express*, June 12, 1996.

Pendleton, George, "John Cabot, a brief account of his life," *The Beaver*, June, 1929.

Piers, Harry, "The Cabots and Their Voyages" in *Canadian History*, Saint John, New Brunswick, June, 1898.

Pierson, Frank W., "Cabotian Literature," *The Newfoundland Quarterly*, Volume 34, Number 4, St. John's, April, 1935.

Poirier, Pascal, "Jean Cabot [re Cabot's Landfall in Newfoundland and Labrador]," *Revue Canadienne*, Volume 43, January & February, 1903.

161

_____ "Jean Cabot" (Extrait de: Le Cap-Breton et Ses Decouveurs) (en voie de preparation), *Revue Canadienne*, Volume 43, January-February, 1903.

Pope, Joseph, "The Cabot Celebration," in the *Canadian Magazine*, Toronto, December, 1896.

Porter, Edward G., "The Cabot Celebrations of 1897," in *The New England Magazine*, Volume 27, Number 6, February, 1898.

Prowse, Daniel W., "The Discovery of Newfoundland by John Cabot in 1497," in the *Royal Gazette*, St. John's, Newfoundland, June-July, 1897.

_____ "Cabot's Landfall," in *The Morning Chronicle*, Halifax, August 7, 1897.

Prowse, George R.F., "Sebastian Cabot Lied," in *Cartological Materials*, Number 3, Winnipeg, 1942.

_____ "The Cabot Landfall," *International Geographical Congress Transactions*, 8th Congress, Volume 8, 1904.

_____ "The Voyages of John Cabot in 1497 to North America" (A *Review* of Henry Harrisse's book, 1896), (mimeographed sheets), Bradford, 1897.

Quinn, David B., "The Croft Voyage of 1481," in *The Mariner's Mirror*, Volume 21, 1935.

_____ "The Argument for the English Discovery of America between 1480 and 1497," *The Geographical Journal*, Volume 127, Part 3, 1961.

_____ "John Cabot's Matthew," *Times Literary Supplement*, June 8, 1967.

Radford, Daniel R., "Newfoundland has pulled off the biggest historical sham of the century," Letter to the Editor, *Downhomer*, Volume 10, Number 6, November, 1997.

Rowe, C.J., "John Cabot," *Atlantic Guardian*, Volume 14, Number 6, June, 1957.

Ruddock, Alwyn A., "The Reputation of Sebastian Cabot," *Bulletin of the Institute of Historical Research*, Volume 47, 1974.

_____ "John Day of Bristol and the English Voyages Across the Atlantic before 1497," *The Geographical Journal*, Volume 132, Part 2, June, 1966.

Rumbolt, Chris, "In search of the real John Cabot," *The Downhomer*, Volume 9, Number 5, October, 1996.

Rumbolt, Curtis, "Who was John Cabot?," *The Express*, St. John's, July 10, 1996.

Ryan, A.N., "Bristol, the Atlantic and North America, 1450-1509," in J. B. Hattendorf (ed.), *Maritime History*, Volume 1, *The Age of Discovery*, Florida, 1966.

Saunders, Robert, "Stand Fast for Newfoundland," *The Newfoundland Quarterly*, Volume 61, Number 4, Winter, 1962.

_____ "Stand Fast for Newfoundland," *The Newfoundland Quarterly*, Volume 61, Number 2, Summer, 1961.

_____ "Cabot's Landfall in the New World," *The Newfoundland Quarterly*, Volume 59, Number 4, St. John's, Winter, 1960.

_____ "Stand Fast for Bonavista," *The Newfoundland Quarterly*, Volume 56, Number 1, St. John's, Spring, 1959.

_____ "Stand Fast for Newfoundland," *The Newfoundland Quarterly*, Volume 50, Number 2, September, 1950.

Selwyn-Brown, Arthur, "Ships of Early Atlantic Voyages," *The Newfoundland Quarterly*, Volume 21, Number 2, St. John's, Spring, 1921.

Sewall, John S., "The Value and Significance of Cabot's Discovery to the World," in the *Proceedings* of the Maine Historical Society, Portland, Me., Volume 8, 1897.

Shaw, Barbara E., "The True History of Giovani Caboto, Grates Cove, and the Cabot Rock," (privately printed) Grates Cove, 1995.

Skelton, Raleigh A., "The Cartography of the Voyages" in James A. Williamson, *The Cabot Voyages and Bristol Discovery under Henry VII*, Cambridge, England, 1962.

_____ "The Cartographic Record of the Discovery of North America," Congresso Internacional de Historia dos Descobriementos, *Actas,* Volume 2, Lisbon, Portugal, 1961. (Reprint, 1965).

_____ "Did Columbus or Cabot See the Map?" *American Heritage,* October, 1965.

_____ "John Cabot" in the *Dictionary of Canadian Biography*, Volume I, Toronto, 1966.

Smith, Roger C., "Vanguard of Empire: The mariners of exploration and discovery," *Terrae Incognitae,* Volume 17, 1985.

Smrz, Jiri, *Cabot and Newfoundland Identity: The 1897 Celebrations*, B.A. (Honours) Thesis, Memorial University, 1994.

Sparks, Carol, "England and the Columbian Discoveries: The attempt to legitimize English voyages to the New World," - *Terrae Incognitae*, Volulme 22, 1990.

Stevens, Henry, "Sebastian Cabot minus John Cabot equals zero," in *The Daily Advertiser*, Boston, March, 1870.

Taylor, Eva, G.R., "Where did the Cabots Go?," *The Geographical Journal*, Volume 129, London, 1963.

Thacher, John Boyd, "The Cabotian Discovery," in the *Proceedings* of the Royal Society of Canada, Volume 3, Ottawa, 1897.

Thorburn, Robert, Michael F. Howley and Moses Harvey, "A Cabot Souvenir Number," St. John's, June, 1897 (with a poem by Isabella W. Rogerson).

True, David O., "Cabot Explorations in North Atlantic," *Imago Mundi*, Volume 13, 1956.

_____ "John Cabot's Maps and Voyages," *Actas 2*, Congresso Internacional de Historia dos Descobrimentos, Lisbon, 1961.

_____ "New Light on the 1492 Voyage of John Cabot," *The Carrell* (Friends of the Miami Library), Volume 1, Number 1, June, 1960.

_____ "Cabot Explorations in North America," *Imago Mundi*, Volume 13, Number 7, 1955.

Vigneras, Louis-Andre, "Etat present des etudes sur Jean Cabot," *Lecture* delivered at the Congress of the History of Discoveries, Lisbon, September, 1960.

_____ "New Light on the 1497 Cabot Voyage to America," *Hispanic-American Historical Review*, Volume 36, Number 4, 1956.

_____ "The Cape Breton Landfall: 1494 or 1497? Note on a Letter by John Day," *Canadian Historical Review*, Volume 38, Number 3, 1957.

Walsh, Pat, "The Legend of the Cabot Rock," *Newfoundland Ancestor*, Summer, 1996.

Whiteley, George, "Was Cape Degrat Cabot's Landfall?," *The Evening Telegram*, St. John's, Newfoundland, June 23, 1984.

_____ "John Cabot's Voyage of Discovery, *The Newfoundland Quarterly*, Volume 72, Number 1, Summer 1986.

Williams, Alan F., "Cabot 500: Myths, Traditions and Celebrations," *Lecture*, Canada House, London, 1996.

Williamson, James A., "John Cabot, the Discoverer of North America," in James A. Williamson (ed.), *Builders of the British Empire*, London, 1925.

Winship, George P., "Some facts about John and Sebastian Cabot," *American Antiquarian Society, Proceedings*, Volume 13, Number 3, 1900.

_____ "John Cabot and the Study of Sources," in the report of *The American Historical Association* for 1897, Washington, D.C., 1898.

_____ "Commentary on Archbishop Cornelius O'Brien's Speech in Halifax, 1897," in *Cabot Bibliography*, New York & London, 1900.

_____ "The Careers of the Cabots," in *Cabot Bibliography*, London, New York, 1900.

Wrong, George M., "The Cabots," in *The Review of Historical Publications Relating to Canada*, Toronto, 1897.

Young, Ewart, "Shades of John Cabot," *Atlantic Guardian,* February 1955.

C: Encyclopedias and Dictionaries

Containing items relative to John and Sebastian Cabot

1. Dictionary of National Biography
2. Dictionary of Canadian Biography (Raleigh A. Skelton)
3. Encyclopedia Britannica
4. Encyclopedia of Canada
5. The Canadian Encyclopedia (John Parsons)
6. Encyclopedia Canadiana
7. Encyclopedia Americana (Raleigh A. Skelton)
8. World Book Encyclopedia (Leslie Harris)
9. Encyclopedia of World Biography
10. Encyclopedia of Newfoundland and Labrador (Pamela M. Hodgson)
11. Dictionary of Newfoundland and Labrador Biography
12. Merit Encyclopedia
13. Compton's Encyclopedia
14. The Marshall Cavendish Illustrated Encyclopedia of Discovery and Exploration
15. The Oxford Companion to Canadian History and Literature (Editor, Norah Storey)
16. Encyclopedia of World History
17. Dictionary of Literary Biography — Volume 47.
    "American Historians, 1866-1912"
    Essay on Henry Harrisse, a famous Cabotian Scholar.
18. Dictionary of Literary Biography — Volume 17.
    "Twentieth Century American Historians."
    Essay on Samuel E. Morison, a famous Cabotian Scholar.
19. Webster's Biographical Dictionary, 1976.
20. The Macmillan Dictionary of Biography, 1985.
21. Funk and Wagnall's New Encyclopedia.
22. Cambridge History of the British Empire.

D: Other Bibliographical Material

Miscellaneous items related to John and Sebastian Cabot can be found in the following:

1. *The Newfoundland Quarterly*, Feb. 1955, Sept. 1954, Sept. 1956, Oct. 1936, Dec. 1955, March 1960, Spring 1962, Summer 1962, March 1959, June 1960, Oct. 1961.

2. *Observer's Weekly*, (Volume 2, Number 3), June 21, 1934

3. Editorial in *The Compass*, "Legend has it," June 4, 1996.

4. Editorial in *The Express*, "Cabot 500th," December 30, 1996.

5. *New-Land Magazine*, Volume 14, Autumn, 1968.

6. *Blackwood's Magazine*, Volume 161, Number 1480, June, 1897.

7. "Report of the Committee appointed in May 1895, in relation to the Commemoration in 1897 of the Discovery of the mainland of North America by John Cabot," in the *Proceedings* of the Royal Society of Canada, Ottawa, October, 1896.

8. *Decks Awash* (Grates Cove), December, 1980.

9. *Newfoundland Churchman*, ("A Gargoyle from Bristol"), November, 1980.

10. *The Evening Telegram*, Supplement, 450th Anniversary of Cabot's Voyage, June 23, 1947.

11. *Jackdaw*, Series, No. 69, (Clarke-Irwin), Toronto, 1967.

12. *Newfoundland Stories and Ballads*, Spring, 1967.

13. *The Downhomer*, October 1996.

14. *The Evening Telegram*, Supplement, 400th Anniversary of Cabot's Voyage, June 24, 1897.

15. *The Compass*, March 25, 1981.

16. *The Evening Telegram,* February 8, 1995.

17. *The Compass,* July 1, 1997 and July 15, 1997.

18. Round the Bay — Cruising Newfoundland and Labrador, 1997.

19. "The Cabot (?) Inscription," *The Daily News,* January 30, 1956.

20. English, Leo E.F., "Letter to the Editor," *The Daily News,* October 20, 1955.

21. A.M. Sullivan's Letter "Cabot and Grates Cove," *Cabot Biographical File,* Centre for Newfoundland Studies, Memorial University of Newfoundland, St. John's.

22. Newfoundland and Labrador *Visitor's Guide,* July/August *1997* Edition.

John Parsons

John Parsons (b. 1939), a native of Shearstown, Conception Bay taught school in Newfoundland for over thirty years. A writer and historian with interests in Newfoundland and Labrador history and culture, and a multiplicity of other subjects, he lives in St. John's, but spends as much time as possible — particularly when he is writing — at his estate in Shearstown bordering on Grassy Pond (Latitude 47⁰-35'N — Longitude 53⁰-19'W). Since his retirement from the teaching profession in 1994, he has devoted his time to research and writing. He is the author of *Labrador: Land of the North* (1970), *The Confederation Movement in Newfoundland 1888-1895* (1972), editor of *All the Luck in the World* (1994) by Colonel Allan M. Ogilvie, co-author of *The King of Baffin Land* (1996), and author of *Away Beyond the Virgin Rocks: A Tribute to John Cabot* (1997). Between 1981 and 1994, Parsons contributed sixty-nine biographical items to the *Encyclopedia of Newfoundland and Labrador,* and during the 1980's he contributed fourteen biographical items to *The Canadian Encyclopedia* (1988). He was also a biography consultant to this encyclopedia. Parsons did graduate studies in history at Memorial University of Newfoundland, and in education at the Ontario Institute for Studies in Education (University of Toronto). John Cabot has occupied his attention now for several years.